200 Original Shop Aids & Jigs for Woodworkers

Rosario Capotosto

Drawings by Michael Capotosto

Popular science

Sterling Publishing Co., Inc.   New York

To Jennie, who was so helpful as usual

Library of Congress Cataloging-in-Publication Data

Capotosto, Rosario.
  200 original shop aids and jigs for woodworkers.

  Originally published as: Capotosto's woodworking
wisdom New York : Popular Science Books : Van Nostrand
Reinhold, c1983.
  Includes index.
  1. Woodwork.   2. Woodworking tools.   I. Title.
II. Title: Two hundred original shop aids and jigs
for woodworkers.
TT180.C364   1987   684'.08   87-10144
ISBN 0-8069-6582-7 (pbk.)

First published in paperback in 1987 by Sterling Publishing Co., Inc.
387 Park Avenue South, New York, N.Y. 10016
Originally published in hardcover by Grolier Book Clubs, Inc.
under the title "Capotosto's Woodworking Wisdom"
Copyright © 1983 by Rosario Capotosto
Distributed in Canada by Sterling Publishing
℅ Canadian Manda Group, P.O. Box 920, Station U
Toronto, Ontario, Canada M8Z 5P9
Distributed in Great Britain and Europe by Cassell PLC
Artillery House, Artillery Row, London SW1P 1RT, England
Distributed in Australia by Capricorn Ltd.
P.O. Box 665, Lane Cove, NSW 2066
Manufactured in the United States of America
All rights reserved

CONTENTS

Appendix

WEIGHTS AND MEASURES

UNIT	ABBREVIATION	EQUIVALENTS IN OTHER UNITS OF SAME SYSTEM	METRIC EQUIVALENT
Weight			
Avoirdupois			
ton			
short ton		20 short hundredweight, 2000 pounds	0.907 metric tons
long ton		20 long hundredweight, 2240 pounds	1.016 metric tons
hundredweight	cwt		
short hundredweight		100 pounds, 0.05 short tons	45.359 kilograms
long hundredweight		112 pounds, 0.05 long tons	50.802 kilograms
pound	lb *or* lb av *also* #	16 ounces, 7000 grains	0.453 kilograms
ounce	oz *or* oz av	16 drams, 437.5 grains	28.349 grams
dram	dr *or* dr av	27.343 grains, 0.0625 ounces	1.771 grams
grain	gr	0.036 drams, 0.002285 ounces	0.0648 grams
Troy			
pound	lb t	12 ounces, 240 pennyweight, 5760 grains	0.373 kilograms
ounce	oz t	20 pennyweight, 480 grains	31.103 grams
pennyweight	dwt *also* pwt	24 grains, 0.05 ounces	1.555 grams
grain	gr	0.042 pennyweight, 0.002083 ounces	0.0648 grams
Apothecaries'			
pound	lb ap	12 ounces, 5760 grains	0.373 kilograms
ounce	oz ap	8 drams, 480 grains	31.103 grams
dram	dr ap	3 scruples, 60 grains	3.887 grams
scruple	s ap	20 grains, 0.333 drams	1.295 grams
grain	gr	0.05 scruples, 0.002083 ounces, 0.0166 drams	0.0648 grams
Capacity			
U.S. Liquid Measure			
gallon	gal	4 quarts (2.31 cubic inches)	3.785 litres
quart	qt	2 pints (57.75 cubic inches)	0.946 litres
pint	pt	4 gills (28.875 cubic inches)	0.473 litres
gill	gi	4 fluidounces (7.218 cubic inches)	118.291 millilitres
fluidounce	fl oz	8 fluidrams (1.804 cubic inches)	29.573 millilitres
fluidram	fl dr	60 minims (0.225 cubic inches)	3.696 millilitres
minim	min	1/60 fluidram (0.003759 cubic inches)	0.061610 millilitres
U.S. Dry Measure			
bushel	bu	4 pecks (2150.42 cubic inches)	35.238 litres
peck	pk	8 quarts (537.605 cubic inches)	8.809 litres
quart	qt	2 pints (67.200 cubic inches)	1.101 litres
pint	pt	½ quart (33.600 cubic inches)	0.550 litres
British Imperial Liquid and Dry Measure			
bushel	bu	4 pecks (2219.36 cubic inches)	0.036 cubic metres
peck	pk	2 gallons (554.84 cubic inches)	0.009 cubic metres
gallon	gal	4 quarts (277.420 cubic inches)	4.545 litres
quart	qt	2 pints (69.355 cubic inches)	1.136 litres
pint	pt	4 gills (34.678 cubic inches)	568.26 cubic centimetres
gill	gi	5 fluidounces (8.669 cubic inches)	142.066 cubic centimetres
fluidounce	fl oz	8 fluidrams (1.7339 cubic inches)	28.416 cubic centimetres
fluidram	fl dr	60 minims (0.216734 cubic inches)	3.5516 cubic centimetres
minim	min	1/60 fluidram (0.003612 cubic inches)	0.059194 cubic centimetres
Length			
mile	mi	5280 feet, 320 rods, 1760 yards	1.609 kilometres
rod	rd	5.50 yards, 16.5 feet	5.029 metres
yard	yd	3 feet, 36 inches	0.914 metres
foot	ft *or* '	12 inches, 0.333 yards	30.480 centimetres
inch	in *or* "	0.083 feet, 0.027 yards	2.540 centimetres
Area			
square mile	sq mi *or* m^2	640 acres, 102,400 square rods	2.590 square kilometres
acre		4840 square yards, 43,560 square feet	0.405 hectares, 4047 square metres
square rod	sq rd *or* rd^2	30.25 square yards, 0.006 acres	25.293 square metres
square yard	sq yd *or* yd^2	1296 square inches, 9 square feet	0.836 square metres
square foot	sq ft *or* ft^2	144 square inches, 0.111 square yards	0.093 square metres
square inch	sq in *or* in^2	0.007 square feet, 0.00077 square yards	6.451 square centimetres
Volume			
cubic yard	cu yd *or* yd^3	27 cubic feet, 46,656 cubic inches	0.765 cubic metres
cubic foot	cu ft *or* ft^3	1728 cubic inches, 0.0370 cubic yards	0.028 cubic metres
cubic inch	cu in *or* in^3	0.00058 cubic feet, 0.000021 cubic yards	16.387 cubic centimetres

METRIC SYSTEM

UNIT	ABBREVIATION		APPROXIMATE U.S. EQUIVALENT
Length			
		Number of Metres	
myriametre	mym	10,000	——————— 6.2 miles
kilometre	km	1000	0.62 mile
hectometre	hm	100	109.36 yards
dekametre	dam	10	32.81 feet
metre	m	1	39.37 inches
decimetre	dm	0.1	3.94 inches
centimetre	cm	0.01	0.39 inch
millimetre	mm	0.001	0.04 inch
Area			
		Number of Square Metres	
square kilometre	sq km *or* km²	1,000,000	0.3861 square miles
hectare	ha	10,000	2.47 acres
are	a	100	119.60 square yards
centare	ca	1	10.76 square feet
square centimetre	sq cm *or* cm²	0.0001	0.155 square inch

UNIT	ABBREVIATION		APPROXIMATE U.S. EQUIVALENT
Volume			
		Number of Cubic Metres	
dekastere	das	10	13.10 cubic yards
stere	s	1	1.31 cubic yards
decistere	ds	0.10	3.53 cubic feet
cubic centimetre	cu cm *or* cm³ *also* cc	0.000001	0.061 cubic inch

UNIT	ABBREVIATION	*Number of Litres*	Cubic	Dry	Liquid
Capacity					
kilolitre	kl	1000	1.31 cubic yards		
hectolitre	hl	100	3.53 cubic feet	2.84 bushels	
dekalitre	dal	10	0.35 cubic foot	1.14 pecks	2.64 gallons
litre	l	1	61.02 cubic inches	0.908 quart	1.057 quarts
decilitre	dl	0.10	6.1 cubic inches	0.18 pint	0.21 pint
centilitre	cl	0.01	0.6 cubic inch		0.338 fluidounce
millilitre	ml	0.001	0.06 cubic inch		0.27 fluidram

UNIT	ABBREVIATION	*Number of Grams*	APPROXIMATE U.S. EQUIVALENT
Mass and Weight			
metric ton	MT *or* t	1,000,000	1.1 tons
quintal	q	100,000	220.46 pounds
kilogram	kg	1,000	2.2046 pounds
hectogram	hg	100	3.527 ounces
dekagram	dag	10	0.353 ounce
gram	g *or* gm	1	0.035 ounce
decigram	dg	0.10	1.543 grains
centigram	cg	0.01	0.154 grain
milligram	mg	0.001	0.015 grain

PREFACE

Books on the subject of woodworking generally are aimed rather specifically at either the beginning or the advanced woodworker. This one is directed toward neither group in particular because its contents will appeal to, and prove helpful to, both the novice who has at least a working familiarity with tools and the more practiced worker.

For example, assuming the use of a table saw but lacking the aid of a tenoning jig, the experienced woodworker conceivably will do a better job of cutting a tricky joint than will the beginner. But with the use of the tenoning jig described in these pages, both can produce the joint with matching quality. With this aid the beginner acquires the capability for perfection while the expert gains the advantage of producing the joint with much greater speed than likely would be possible with his or her usual, probably makeshift, methods.

The particular jig cited is somewhat elaborate and does rely on a certain degree of know-how with power tools to construct. Does this leave the beginner thwarted? Not at all. The easy-to-follow construction plan and text as well as the how-to photographs describe clearly how to build it, step by step.

The section on techniques will similarly prove useful to all woodworkers. The ideas range wide and some are rather simple—but all will help the reader to get more out of woodworking more easily. The practical projects in the last section of the book provide a pleasurable means for further development of techniques.

It should be noted that in some photographs involving power tools the safety guards have been removed for the purpose of clarity of an operation. In actual application, *make use of all guards and safety devices on tools that are so equipped.*

Some of the material in this book has previously appeared in magazines including *The Family Handyman* and *Popular Mechanics*. Changes have been made in the presentation of the material to suit the format of this book.

Thanks to Michael Capotosto for making the drawings and thanks also to my editor, Henry Gross, for his assistance in the preparation of this book, and to Andrew Steigmeier and Jeff Fitschen for their design and layout.

R.C.

PART I

Making and Using Jigs and Aids

IN THE COURSE OF many woodworking activities there frequently arises the necessity to find a better and perhaps a simpler way to perform an operation. Thus various jigs have evolved—usually homemade gadgets or contrivances that help to get the job done expeditiously. They may serve to ensure accuracy, to speed the work, or to obtain consistent results in producing similar parts.

On these pages you will find a collection of useful jigs and aids that you can make; these will help you to achieve greater precision and to perform certain operations with ease and increased speed.

Also included is a chapter on metalworking jigs. This may seem to be out of place in a book dealing with woodwork, but it definitely does belong. Woodwork projects quite often include metal components. You will be ahead in both satisfaction and thriftness if you can custom-fabricate pieces rather than choose to have the metalwork done for you, usually at considerable expense, by a professional. The jigs shown in the metalworking chapter concentrate on forming metal because this is the main stumbling block for many home workshoppers. The specific end-products are incidental; you can adapt the basics for your own particular applications.

1. HANDSAW

Dovetail Guide

Dovetail joints of good quality demand precise work with the saw and chisel. Start off right by using this handy jig to make the initial cheek cuts accurately. The depth of the notch in the block should be sized so the depth of the cut equals the thickness of the workpiece. A companion jig with a horizontally angled notch is used to guide the saw for the mating socket cuts.

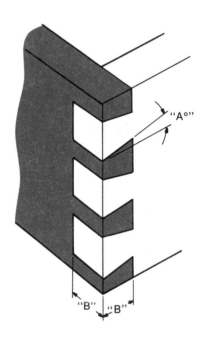

"A°" = DOVETAIL PIN ANGLE
"B" = THICKNESS OF MATERIAL

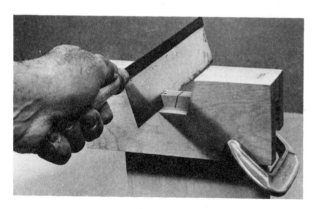

Saw is guided against the face of the block to make the angled cut. Block is alternately shifted to make the opposing angle cut. Jig can be used for dovetails of varying width.

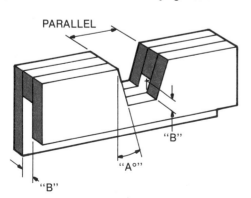

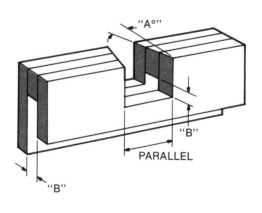

Crown Molding Miter Box

The installation of crown or hollow-back cove moldings with mitered inside or outside corners necessitates compound angle cuts—a combination of a bevel and miter—because this type of molding is designed to be installed in a slanted position. Such a cut is practically impossible to make freehand due to the difficulty of accurately marking the cutting lines and following them with the saw.

The problem is solved by making a miter box with side walls spaced to hold the molding at its normal, slanted installation position. Determine the width that holds the molding at the proper angle, assemble the box, and carefully make a left-and-right perpendicular 45-degree mitered kerf cut.

Straight miter cuts in the molding with this jig will automatically result in the correct compound bevel-miter angle for a well fitted joint.

Strip of tape is used to keep the work from shifting sideways. Backsaw is best for this work, but if one is not available the job can be done with a regular, fine-tooth handsaw.

Half-Lap Guide

Two strips of wood half the thickness of the workpiece serve as depth-of-cut guides in this simple jig. Clamp the strips along each side of the work to hold it firmly in place; then make a series of closely spaced kerf cuts, sawing until the teeth reach the guides. Use a wide chisel to break off the slices. This procedure results in greater accuracy than would be obtained in attempting to make the face cuts by lengthwise ripping. The same setup is used for cutting middle half-laps that would otherwise be impossible with the handsaw. Tenon cuts can also be guided in a similar manner by simply varying the thickness of the guides.

Saw is held as level as possible so the teeth will strike both guides at the same time.

Coping Saw V Block

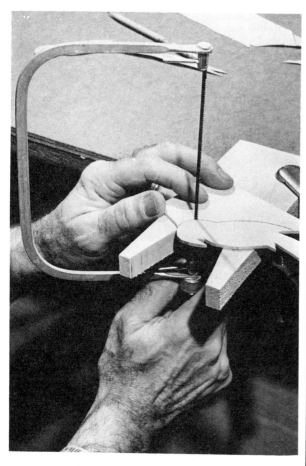

Board with a V cutout is clamped to the workbench and provides saw clearance and a chatter-free solid support for the work. Blade teeth point down toward the handle for this cutting method, which affords good visibility of the cutting line.

Saw Jointing Jig

When the teeth of a handsaw blade have been worn unevenly by use or abuse, the first step in the sharpening sequence is to bring all the teeth to the same height. This file-holding and guiding jig will enable you to joint the teeth safely and accurately. Space the side blocks for a slide fit over the blade and be sure to bevel the upper inside corners to allow clearance for the set teeth.

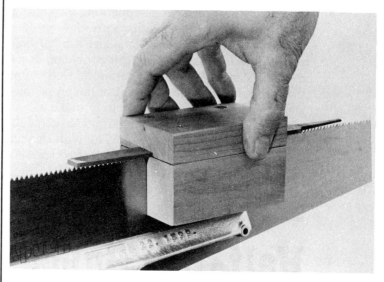

Jig is passed lightly over the teeth until all teeth are filed to a uniform height. Several passes usually will do.

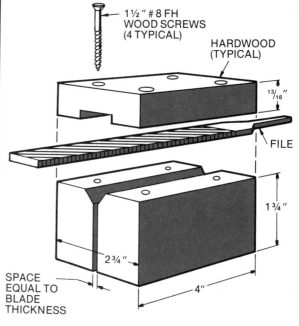

1½" #8 FH WOOD SCREWS (4 TYPICAL)

HARDWOOD (TYPICAL)

13/16"

FILE

1¾"

2¾"

4"

SPACE EQUAL TO BLADE THICKNESS

2. SABRE SAW

Notched Joint Jig

The notch for the notch-and-tenon joint used in the construction of the bunk bed shown here is best accomplished with the sabre saw. Although the table saw can be used to make the right-angle cuts that form the tenon, it cannot be used to make the notch cutout. Here you need a sabre saw with this simple jig. If you do not have access to a table saw, all the cuts can be made with the sabre saw.

Notch-and-tenon joint, which exposes contrasting woods, is a design feature of this piece. Sabre saw is the most practical tool to use for making the notch cuts.

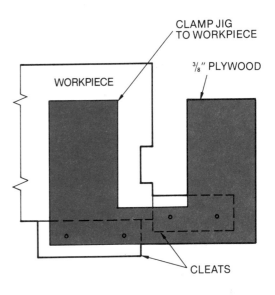

The jig consists of a piece of ³/₈-inch plywood with a rectangular cutout. It serves to guide the saw straight for the inside cut and also stops the forward travel of the saw at the end of the cut. The size of the cutout in the jig is variable depending on the size of your saw's base. Make the cutout about an inch or so wider than the width of the saw base.

All workpieces must be clearly marked to avoid errors. The waste areas are indicated by X's.

Two cleats nailed to a workboard that is clamped to the drill press table serve as stops to ensure consistency in locating the blade access holes.

Saw rides against the edge of the cutout and stops when its base front reaches the front edge of the cutout. Cleats position the jig; a clamp holds it in place.

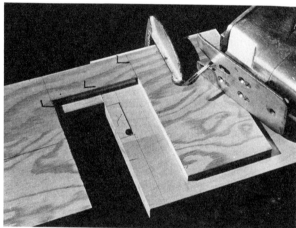

Close-up shows how the jig cutout is offset from the notch cutting lines.

Two cleats are tack nailed to butt against the workpiece edges and ensure uniform positioning of the jig on all the pieces to be cut. To position the jig, the front and side edges of the cutout are offset from the cutting lines an amount equal to the distance from the front and side of the blade to the front and side of the saw base, respectively.

The cut is started by placing the blade into the entry hole bored in one corner. When the long cuts have been made in all the pieces the jig is repositioned (and cleats relocated) to make the short, endwise cuts. These cuts can also be made on the table saw. If the table saw is used, set the blade high to minimize the depth of the kerf undercut on the back side of the work.

It should be noted that in sabre-saw work the good, or face, side of the stock should face down; however, this rule can be waived provided a very smooth-cutting, fine-tooth blade is used to avoid splintering the face edges. Another alternative is to use a "teeth-down" blade, which cuts on the downstroke.

Short, endwise cuts can be cut on the table saw. High blade results in a shallow kerf on the underside. Sabre saw and jig will be used to make the crosscuts for this tenon because it is rather difficult to guide long workpieces accurately with a miter gauge.

Unit is dry-assembled to allow accurate drilling of the screw pilot holes. Larger counterbore holes for the plugs are drilled in advance.

Joint ends should project slightly to allow flush sanding to the veneer surfaces. Portable belt sander works best for the preliminary stock removal.

Jig for Edge Half-Laps

A jig similar to the one above can be used to make the notch cuts for the edge half-lap joint. The only difference in the jig is that the cutout must be precisely sized. Make the width of the opening equal to the width of the saw base, plus the thickness of the stock, minus the thickness of the blade.

Two cuts alternately made with the saw base guided along each edge of this carefully sized cutout result in an accurately sized notch when the waste is dropped out.

Table Mount

This jig readily converts the sabre saw into a stationary scrolling saw. This is especially advantageous for handling small workpieces. An adjustable sliding arm with pivot makes quick and easy work of cutting perfectly circular discs. The saw is mounted with four screws and can be removed quickly for regular service when desired.

MATERIALS LIST

ITEM	QTY	DESCRIPTION
1	2	$1/2'' \times 14'' \times 14''$ plywood
2	4	$2 \times 4 \times 7^{1}/4''$
3	2	$3/4'' \times 3^{1}/2'' \times 20''$ pine
4	1	$1/2'' \times 1^{7}/8'' \times 13^{1}/2''$ oak
5	1	$1/4'' \times 1^{1}/2''$ thumbscrew
6	1	$1/4''$ tee nut
7	9	2" finishing nail
8	8	2" common nail

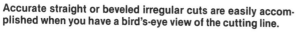

Accurate straight or beveled irregular cuts are easily accomplished when you have a bird's-eye view of the cutting line.

Using the circle cutting accessory. It can be used to cut a disc up to 32 inches in diameter.

Four holes are drilled in the saw base to permit mounting with round head screws.

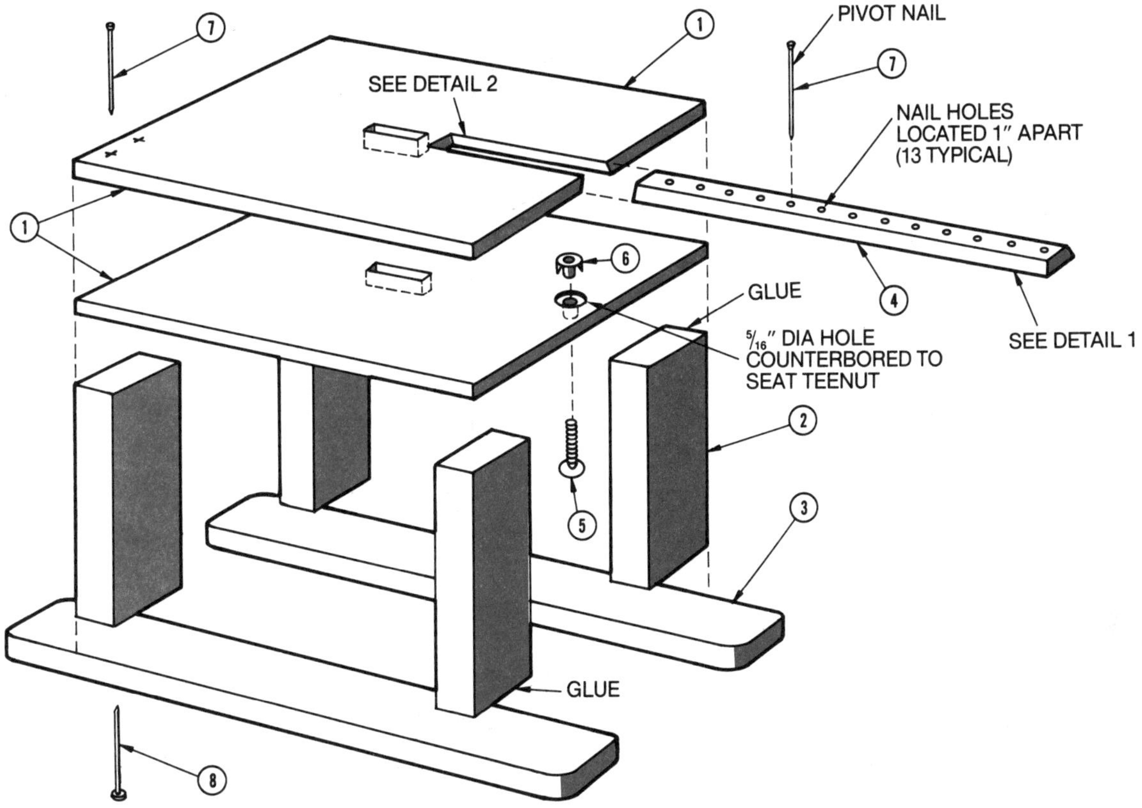

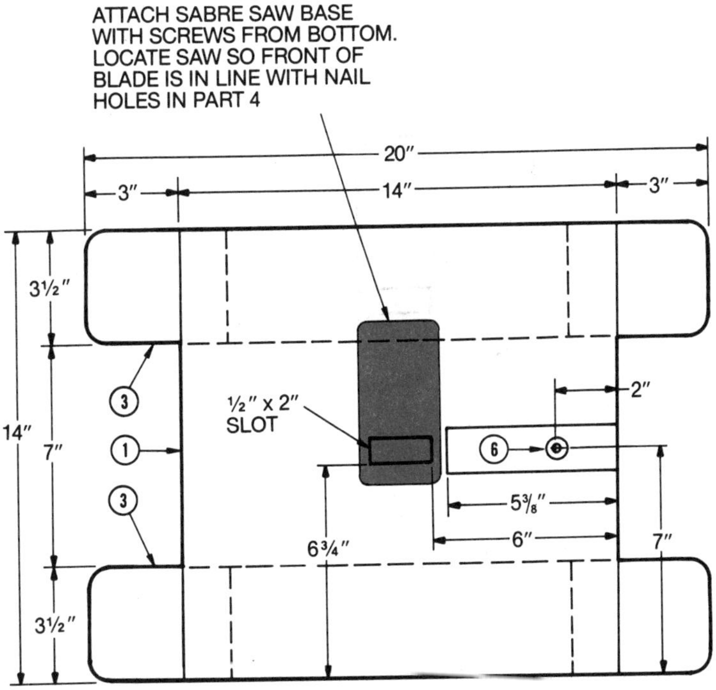

⑦

① SEE DETAIL 2

PIVOT NAIL

⑦

NAIL HOLES
LOCATED 1" APART
(13 TYPICAL)

①

⑥

GLUE

④

5/16" DIA HOLE
COUNTERBORED TO
SEAT TEENUT

SEE DETAIL 1

②

③

⑤

GLUE

⑧

ATTACH SABRE SAW BASE
WITH SCREWS FROM BOTTOM.
LOCATE SAW SO FRONT OF
BLADE IS IN LINE WITH NAIL
HOLES IN PART 4

20"

3"   14"   3"

3½"

③

14"   7"   ①

③

½" x 2"
SLOT

⑥

3½"

6¾"

5⅜"

6"

2"

7"

PLAN VIEW

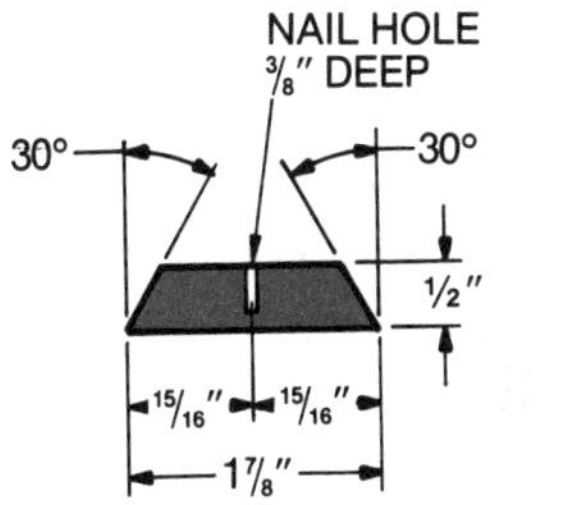

NAIL HOLE
³⁄₈" DEEP

30°   30°

½"

¹⁵⁄₁₆"   ¹⁵⁄₁₆"

1⅞"

DETAIL 1

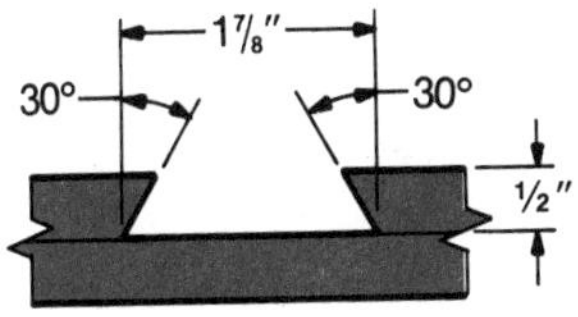

1⅞"

30°   30°

½"

DETAIL 2

3. PORTABLE CIRCULAR SAW

Crosscut Platform

Make precise right-angle or left-and-right miter cuts up to 45 degrees in work up to 7/8 inch thick by 12 inches wide with this jig. Since the saw is not locked into the track but merely placed onto it for use, the setup takes only a few seconds; thus the saw is always ready for other use.

The jig is sized for the 5½-inch Skil Trim Saw, but the track assembly can be altered in size in order to suit any saw. If a saw of larger diameter is to be accommodated, the height of the track assembly can be increased. The result will be a greater thickness cutting capacity.

A bolt "pin" dropped into one of the three guide holes will lock the track accurately for 90-degree right-angle cuts or for left and right 45-degree miter cuts. A small C clamp is used to lock the track for any in-between miter angle cuts.

It should be noted that continued passing of the saw over the upper portion of the pivot dowel will eventually wear it away at the top, but this is of no consequence because the dowel remains intact in the baseboard.

Using the crosscut platform jig to make miter cuts. The pin stop ensures quick and exact adjustment for miters and right-angle cuts.

MATERIALS LIST

ITEM	QTY	DESCRIPTION
1	1	5/8" × 20" × 60" particleboard
2	1	3/4" × 2" × 60" pine
3	2	3/4" × 2½" × 24½" pine
4	2	3/4" × 2½" × 16" pine
5	2	3/4" × 2½" × 7" pine
6	1	3/4" × 7 5/8" × 34" pine
7	2	1 3/8" × 1 9/16" × 7 5/8" fir
8	1	3/4" × 3 1/4" × 34" pine
9	1	3/4" × 2½" × 34" pine
10	1	1/2" × 1" × 34" pine
11	1	1/2" × 1 1/8" × 34" pine
12	1	1/8" × 1 3/4" × 3" hardwood
13	1	3/4" × 1 7/8" dowel
14	1	1" × 3" × 3" fir
15	2	3/8" × 4" hex-hd bolt
16	4	3/8" flat washer
17	2	3/8" hex-hd nut
18	1	3/8" × 3" hex-hd bolt
19	4	1½" # 8 fh wood screw
20	6	2" common nail
21	6	1½" finishing nail
22	36	1¼" finishing nail

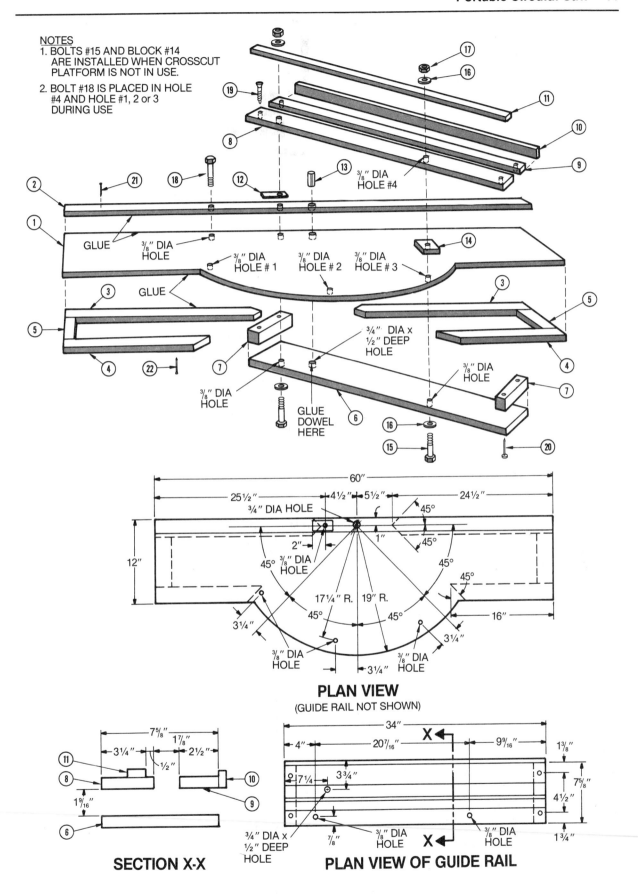

NOTES
1. BOLTS #15 AND BLOCK #14 ARE INSTALLED WHEN CROSSCUT PLATFORM IS NOT IN USE.
2. BOLT #18 IS PLACED IN HOLE #4 AND HOLE #1, 2 or 3 DURING USE

GLUE

³⁄₈" DIA HOLE

³⁄₈" DIA HOLE #4

³⁄₈" DIA HOLE # 1

³⁄₈" DIA HOLE # 2

³⁄₈" DIA HOLE # 3

GLUE

³⁄₄" DIA x ½" DEEP HOLE

³⁄₈" DIA HOLE

³⁄₈" DIA HOLE

GLUE DOWEL HERE

PLAN VIEW
(GUIDE RAIL NOT SHOWN)

60"
25½"
4½"
5½"
24½"
³⁄₄" DIA HOLE
45°
2"
1"
45°
12"
45°   ³⁄₈" DIA HOLE
45°
45°
17¼" R.   19" R.
45°
45°
³⁄₈" DIA HOLE
3¼"
³⁄₈" DIA HOLE
3¼"
16"
3¼"
³⁄₈" DIA HOLE

SECTION X-X

7⁵⁄₈"
1⁷⁄₈"
3¼"
2½"
½"
1⁹⁄₁₆"

PLAN VIEW OF GUIDE RAIL

34"
4"
20⁷⁄₁₆"
9⁹⁄₁₆"
1³⁄₈"
7¼
3¾"
7⁵⁄₈"
4½"
³⁄₄" DIA x ½" DEEP HOLE
⁷⁄₈"
³⁄₈" DIA HOLE
³⁄₈" DIA HOLE
1³⁄₄"

Saw base is set in place to obtain proper spacing of the second track strip. Glue is not applied until the saw is passed through to check for snug, smooth sliding.

Jig can be stored conveniently when not in use. Two bolts secure the track assembly. A spacer block is inserted between the track and base to prevent undue twisting during storage.

Track assembly is clamped in place while the holes for the stop pin are bored.

Quick-Set Rip Guide

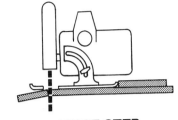

FIRST STEP

The fuss and bother of back-measuring for the base offset factor will be a thing of the past when you use this straightedge guide. The offset is built into the guide so in use its edge is placed directly on the cutting line.

The guide is made by attaching a narrow strip to a piece of plywood that is wide enough to extend an inch or so beyond the free side of the saw base. With the narrow strip serving as the saw guide, a cut is made to slice off the excess. The new edge thus aligns precisely with the inside edge of the saw blade.

It should be noted that the guide will result in accuracy only when you use a blade with the same thickness as the one used to make the initial cutoff.

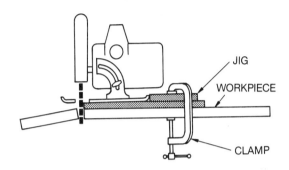

JIG

WORKPIECE

CLAMP

SECOND STEP

Expensive cutting errors are not likely when you use this guide.

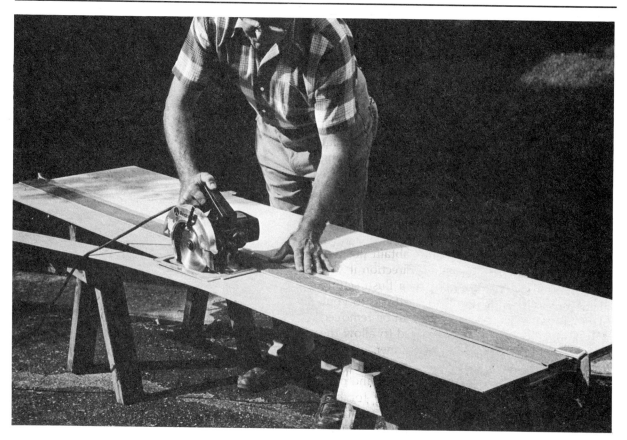

Cut is made to rip the excess from the guide's baseboard.

Workpiece is marked exactly where the cut is desired. No allowance need be made for the blade-to-base-edge offset.

Edge of the guide is placed directly on the mark. The cut will then be precise.

Sawhorse Panel Platform

Sawing large plywood panels on sawhorses requires proper support to prevent the work from caving in and binding against the blade as the cut progresses or to prevent the cutoff from falling away as the cut is completed. This usually involves fumbling with several 2 × 4s to place them in a proper supporting position. And when a second crosswise cut is to be made the supports must be shifted accordingly.

You can avoid the bother and obtain positive support regardless of the sawing direction if you make this pair of sawhorses with a flush setting platform. Or you can adapt the platform to your present sawhorses. The platform can be removed quickly and stored when not needed to allow the sawhorses to be used in the usual manner.

You will need six 8-foot lengths and one 4-foot length of 2 × 4 stock to make the horses and platform. Select straight, kiln-dried lumber for best results. The leg tops require steep angle cuts. The radial-arm saw is ideally suited for this. You can

This arrangement lets you saw in any direction without shifting the supports.

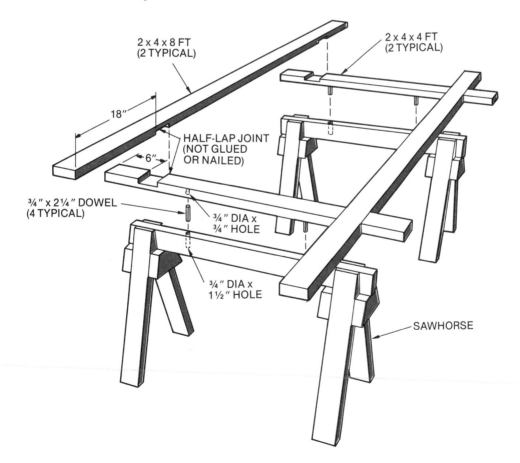

2 x 4 x 8 FT
(2 TYPICAL)

2 x 4 x 4 FT
(2 TYPICAL)

18"

HALF-LAP JOINT
(NOT GLUED
OR NAILED)

6"

¾" x 2¼" DOWEL
(4 TYPICAL)

¾" DIA x
¾" HOLE

¾" DIA x
1½" HOLE

SAWHORSE

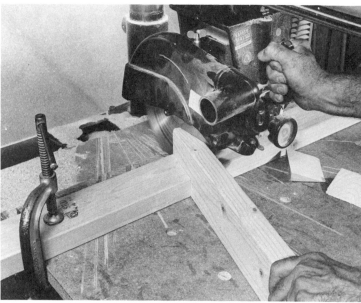

Two mitered pieces of stock are clamped to the saw table to support the sawhorse leg firmly at the correct angle for the steep miter cut.

Saw is set at the normal right-angle position for this cut. The alternative is to swing the saw with the work held perpendicular to the fence.

Platform, ready for use.

swing the arm to 18 degrees and position the work perpendicular to the fence. Or you can keep it at zero and employ the method shown here, in which you miter the ends of two pieces of stock 18 degrees and clamp them to the table, spaced to nest the workpiece at the required 72-degree angle to the saw fence. The mitered support pieces are not wasted; they are later cut down for use as the leg braces. Regardless of which method you select to cut the miters, the clamped 2 × 4s are essential to hold the work steady and safely.

The bottoms of the legs are also mitered 18 degrees so they will rest flush on the floor. Tilt the blade to make these cuts. Use 2-inch hot-dipped galvanized nails and glue to attach the legs.

Use a dado head or a plain crosscut blade to cut the half-lap notches for the platform members. Adjust the blade height so it cuts through half the thickness of the stock. Center the shorter platform pieces on the sawhorse tops and drill centering pilot holes through into the sawhorse tops. Remove the pieces and bore the blind holes for the dowels, as shown in the plan. Glue the dowels only into the bottoms of the cross members. Sand the projecting dowels so they will fit into the mating holes without forcing. Bevel the ends of the dowels so they will find their way into the holes more easily.

When sawing on the platform, adjust the blade projection so it cuts about $1/8$ inch below the surface of the work.

Simple Half-Lap Guide

Jig is used to guide the saw for a series of kerf cuts half the depth of the stock to form half-lap joint notches. Chisel is used to break off the waste slices.

Two sticks nailed together to form a 90-degree right-angle cross probably constitute the simplest jig you can make for your portable saw. The top member of the cross should be equal in thickness to the depth of the desired cut (the jig is shown up side down here). Use the jig to guide the saw for a series of kerf cuts half the depth of the stock to form half-lap joint notches. A chisel is used to break off the waste slices.

Crosscut Guide

Accurate crosscuts, straight or mitered, are easy with this handy guide. One arm is fixed for 90-degree cuts; the other is adjustable, permitting mitered cuts up to 45 degrees. The large handle provides a firm, steady grip.

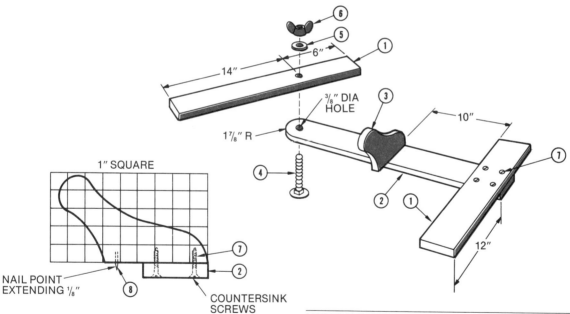

PART 3

1″ SQUARE

NAIL POINT EXTENDING ⅛″

COUNTERSINK SCREWS

¾″ DIA HOLE

1⅞″ R

14″

6″

10″

12″

MATERIALS LIST		
ITEM	QTY	DESCRIPTION
1	2	¾″ × 3″ × 20″ plywood
2	1	¾″ × 3¾″ × 23″ plywood
3	1	¾″ × 5″ × 9″ plywood
4	1	⅜″ × 2″ carriage bolt
5	1	⅜″ flat washer
6	1	⅜″ wing nut
7	6	1½″ # 10 fh wood screw
8	1	1½″ nail (cut to ¾″)

4. SCROLL SAW

Rip and Crosscut Guide

The scroll saw is unsurpassed for the freehand cutting of irregular shapes, but it can also be utilized for ripping and crosscutting with excellent results. The two-component guide shown here will enable you to do precision straight-line cutting on your scroll saw. The rip fence is the basic unit and can be used independently. The crosscut guide mounts onto the rip fence and rides it piggyback fashion.

The guide will permit handling work of unlimited length on scroll saws that feature a table and

Two thumbscrews threaded into tee nuts bear against the bottom of the track to lock the rip fence in place. A scribed centerline on the track, in line with the blade, is used to adjust fence for width of cut.

Oak is a good choice for the guides, but any hardwood will do. The small notch visible in the crosscut guide on the right allows for blade clearance.

Side approach permits ripping stock of unlimited length. A coarse, wide blade is best for ripping. A fast blade stroke and slow feed will produce relatively smooth edges.

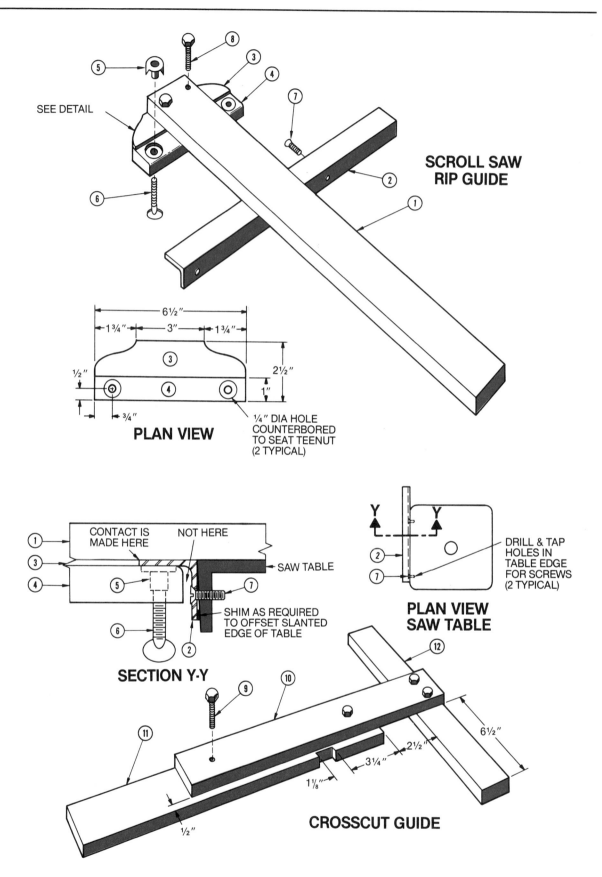

SEE DETAIL

SCROLL SAW RIP GUIDE

6½″
1¾″   3″   1¾″
½″
③
④
2½″
1″
¾″
¼″ DIA HOLE COUNTERBORED TO SEAT TEENUT (2 TYPICAL)

PLAN VIEW

CONTACT IS MADE HERE   NOT HERE
SAW TABLE
SHIM AS REQUIRED TO OFFSET SLANTED EDGE OF TABLE

SECTION Y-Y

Y   Y
DRILL & TAP HOLES IN TABLE EDGE FOR SCREWS (2 TYPICAL)

PLAN VIEW SAW TABLE

6½″
2½″
3¼″
1⅛″
½″

CROSSCUT GUIDE

Note how the crosscut guide fits over the rip fence. Fit must be snug, but not binding. Waxed surfaces prevent binding.

Making a crosscut. Note that only a small, inside cutoff is possible when the workpiece is resting on the table.

blade chucks that can be rotated 90 degrees. For ripping without interference from the overarm, the blade is mounted with the teeth pointing to the left. In this mode the maximum inside crosscut cutoff will be limited by the overarm distance. However, by returning the blade to the normal, front-facing position and rotating the table 90 degrees counterclockwise, stock of unlimited length can be crosscut without any difficulty.

The rip fence travels on an angle iron track that is mounted on the left side of the table with two screws driven into drilled and tapped holes. Or bore track mounting holes at the front of the table. By attaching the track at the front, the table need not be rotated when crosscutting oversized work.

The track must be mounted so its top face is flush and on a true plane with the table top. If the side wall of the table is not perpendicular, as is usually the case with nonmachined castings, use shims to get it right. When the track has been installed, use a square and a metal scriber to scratch a centerline on its face in line with the blade. This will serve as a guide for setting the rip fence for the desired width of cut.

Make the guides of hardwood. Complete the rip fence first, then assemble the crosscut guide so it will travel on the rip fence with a snug, slip fit. Seal all surfaces with several thin coats of shellac. Sand after each coat. Apply a paste wax and buff well for a final finish.

RIP AND CROSSCUT GUIDE
A. RIP FENCE
MATERIALS LIST

ITEM	QTY	DESCRIPTION
1	1	$^{13}/_{16}" \times 2^{1}/_{2}" \times 20"$ hardwood
2	1	$^{1}/_{8}" \times 1^{1}/_{4}" \times 1^{1}/_{4}" \times 16"$ angle iron
3	1	$^{1}/_{8}" \times 1^{1}/_{2}" \times 6^{1}/_{2}"$ hardboard
4	1	$^{13}/_{16}" \times 2^{1}/_{2}" \times 6^{1}/_{2}"$ hardwood
5	2	$^{1}/_{4}"$ tee nut
6	2	$^{1}/_{4}" \times 1^{1}/_{2}"$ thumbscrew
7	2	10–32 $\times$ $^{3}/_{4}"$ fh machine screw
8	2	$^{1}/_{4}" \times 1^{1}/_{4}"$ lag screw

B. CROSSCUT GUIDE

9	4	$^{1}/_{4}" \times 1^{1}/_{4}"$ lag screw
10	1	$^{13}/_{16}" \times 2^{1}/_{2}" \times 16^{1}/_{2}"$ hardwood
11	1	$^{13}/_{16}" \times 3" \times 18"$ hardwood
12	1	$^{13}/_{16}" \times 2" \times 15"$ hardwood

Increased cutoff capacity is gained by resting the work on the ledge of the crosscut guide. A scrap of wood is placed between the table and the stock at the opposite end to support the work on an even plane.

Overarm Circle Jig

This novel circle-cutting jig is designed to mount in either one of the two blade guide post brackets found on the Rockwell scroll saw. With minor modification of the attachment method, the jig can be adapted to other saws. As sized and mounted here, the circle diameter capacity is adjustable up to 18 inches.

The post is made by filing or grinding a flat section on one end of a threaded steel rod. The adjustable pivot is made by grinding a point on the end of a threaded steel rod or a bolt with the head cut off. An easy way to grind a centered point is by chucking the bolt in a drill press or lathe and stroking the end with a file. Another method is to chuck the bolt in a portable drill, then hold the spinning end against a spinning grinding wheel.

This simple jig can be made with a few nuts and bolts and a scrap of wood.

MATERIALS LIST		
ITEM	QTY	DESCRIPTION
1	1	$^{13}/_{16}'' \times 2^{1}/_{2}'' \times 9^{3}/_{4}''$ hardwood
2	1	$^{1}/_{4}'' \times 3''$ threaded steel rod
3	1	$^{1}/_{4}''$ wing nut
4	2	$^{1}/_{4}''$ flat washer
5	1	$^{1}/_{4}''$ hex nut
6	1	$^{1}/_{2}'' \times 6''$ threaded steel rod
7	1	$^{1}/_{2}''$ hex nut
8	1	$^{1}/_{2}''$ hex lock nut

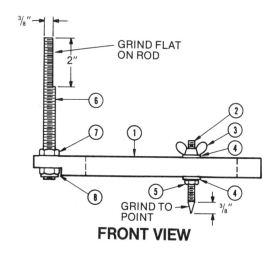

FRONT VIEW

PLAN VIEW
PART 1

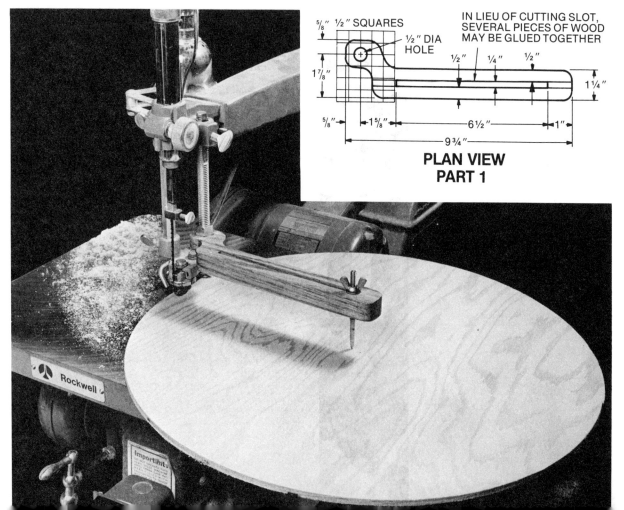

Super-Size Auxiliary Table

The typical home workshop scroll saw has a table that is of ample size to support average-sized work, but large workpieces that overhang the table are difficult to handle. The effort required to support the work puts undue strain on the hands and arms, and this can result in erratic workmanship.

The solution is to make a large auxiliary table such as the one shown here. It will enable you to concentrate your efforts on precision cutting rather than on struggling to balance the workpiece.

The table includes a circle-cutting fixture with a capacity to cut true circles up to 48 inches in diameter on a saw with a 24-inch throat (such as this one). The pivot point is mounted on a sliding bar which can be adjusted for cutting a variable range of circular diameters. The table slides into position on two tracks. This is an important design feature for circle cutting because it effects a tangent lead-in cut from any location on the workpiece. Thus, unlike conventional pivot circle guides, which necessitate

The big table lets you handle large workpieces with relative ease and permits concentration on accuracy.

starting the cut from an outside edge or starting by inserting the blade through a hole bored on the cutting line, this system cuts a disc that is free from a flat or a circular starting indent.

Here is how the sliding table functions in making a circular cut: the work is placed on the prepositioned pivot with the table pulled back. The power is turned on, and the table is pushed forward to engage the work with the blade. When the table comes to a stop the work is rotated to make the circular cut.

Pivoting cuts will track in a true circle only when the pivot and the front edge of the blade are in exact right-angle alignment. Therefore, the pivot must have the capability to be shifted forward or backward in order to align with blades of different widths. Provision is made for this adjustment by setting the eccentric table stop, which is located on the back side of the table.

The table is sized for the Rockwell 24-inch scroll saw. Alter the dimensions as required to suit other saws. Begin construction with two pieces of plywood, each 5/8 by 36 by 36 inches. Make the cuts required but *do not* make the circular cut until the two panels have been glued together. This will preserve the straight edges for accurate measuring and marking. The pivot clearance slot is cut after the recess for the pivot bar has been routed. The blade clearance slot is cut after the two panels have been assembled. The circular cut is made last.

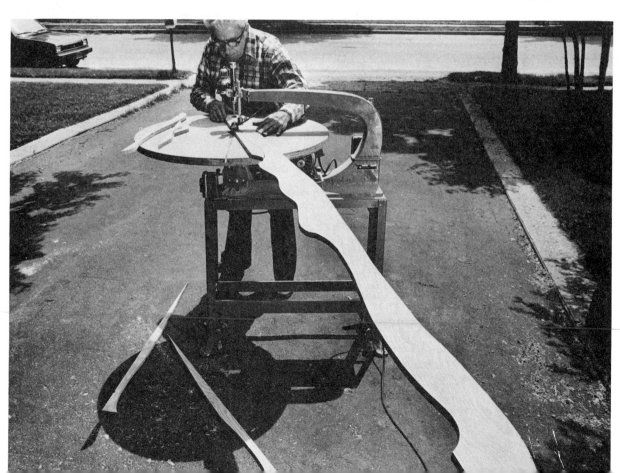

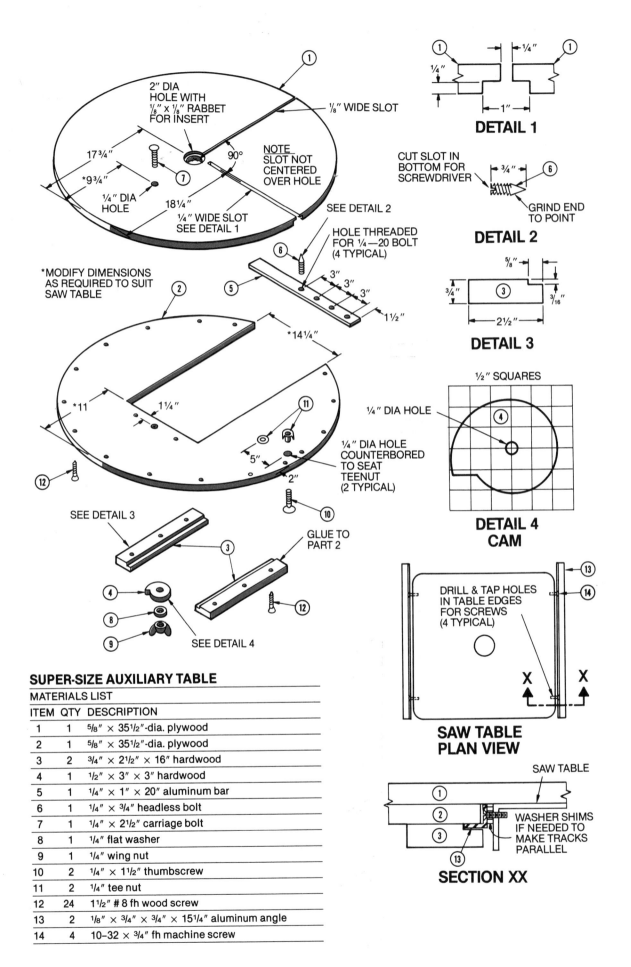

DETAIL 1

CUT SLOT IN
BOTTOM FOR
SCREWDRIVER

GRIND END
TO POINT

DETAIL 2

DETAIL 3

½" SQUARES

¼" DIA HOLE

**DETAIL 4
CAM**

2" DIA
HOLE WITH
⅛" x ⅛" RABBET
FOR INSERT

⅛" WIDE SLOT

17¾"

*9¾"

¼" DIA
HOLE

90°

NOTE
SLOT NOT
CENTERED
OVER HOLE

18¼"

¼" WIDE SLOT
SEE DETAIL 1

SEE DETAIL 2

HOLE THREADED
FOR ¼—20 BOLT
(4 TYPICAL)

*MODIFY DIMENSIONS
AS REQUIRED TO SUIT
SAW TABLE

3"
3"
3"

1½"

*14¼"

*11

1¼"

5"

2"

¼" DIA HOLE
COUNTERBORED
TO SEAT
TEENUT
(2 TYPICAL)

SEE DETAIL 3

GLUE TO
PART 2

SEE DETAIL 4

DRILL & TAP HOLES
IN TABLE EDGES
FOR SCREWS
(4 TYPICAL)

X   X

**SAW TABLE
PLAN VIEW**

SAW TABLE

WASHER SHIMS
IF NEEDED TO
MAKE TRACKS
PARALLEL

SECTION XX

SUPER-SIZE AUXILIARY TABLE

MATERIALS LIST

ITEM	QTY	DESCRIPTION
1	1	⅝" × 35½"-dia. plywood
2	1	⅝" × 35½"-dia. plywood
3	2	¾" × 2½" × 16" hardwood
4	1	½" × 3" × 3" hardwood
5	1	¼" × 1" × 20" aluminum bar
6	1	¼" × ¾" headless bolt
7	1	¼" × 2½" carriage bolt
8	1	¼" flat washer
9	1	¼" wing nut
10	2	¼" × 1½" thumbscrew
11	2	¼" tee nut
12	24	1½" # 8 fh wood screw
13	2	⅛" × ¾" × ¾" × 15¼" aluminum angle
14	4	10-32 × ¾" fh machine screw

The tracks are made of two pieces of $^1/_8$-by-$^3/_4$-by-$^3/_4$-inch aluminum angle. They are attached with two screws driven into holes drilled and tapped in the left and right sides of the table.

A $^1/_4$–20 bolt with head cut off and ground to a point serves as the pivot (see Overarm Circle Jig, above, for details on how to grind the point). The holes in the aluminum bar are tapped to receive the pivot. Notice that when cutting large circles on large workpieces that locate the pivot beyond the table, a few washers placed over the pivot will support the work on the same plane as the table top.

Router with a mortising bit is used to cut the recess for the pivot bar. Tack nailed strips guide the router.

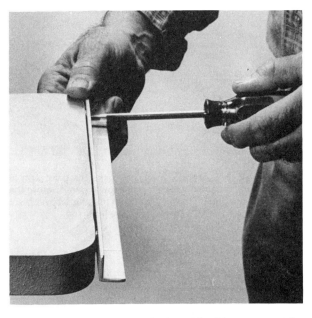

Holes are drilled and tapped in the table sides to mount the aluminum angle tracks. If the table sides are irregular, washer shims are used in order to get the two tracks parallel.

Hardwood track runners must be installed for a slip fit. The table is removed from the saw to facilitate accurate installation.

Worm's-eye view of table as it is moved into place on the tracks. The eccentric stop butts against the front of the saw table. It can be adjusted so the table's pivot will be in alignment with blades of varying widths.

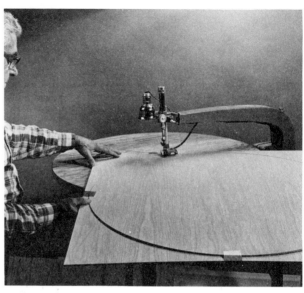

Thumbscrew threaded into a tee nut locks the pivot arm in place. Tee nut must be installed in the lower panel before the two panels are glued together. Washers are placed over pivot so the work will rest at same level as the table top.

Work is rotated when the table comes to a full stop. Notice how tape is used to keep the waste in place.

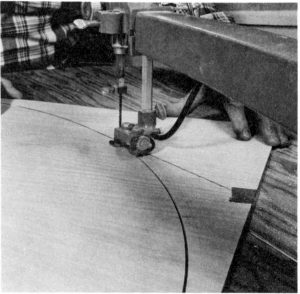

Retracted table and workpiece are pushed ahead into the blade to start a circular cut. A small hole in the bottom of the workpiece is seated on the pivot point. With this arrangement the lead-in cut can be started at any location.

Nearing the end of the cut. Notice how the straight lead-in cut is tangent to the circle. If the blade and pivot are in exact lateral alignment, the cut will be so perfect that it is almost impossible to determine where the cut began and ended on the disc.

Curved Segment Jig

Curved segments are easy to cut with this Y-shaped jig. It is mounted on the arm of the auxiliary table by substituting a longer but unpointed headless bolt for the pointed pivot.

To use the jig, the ends of the arm are tack nailed to the work. The resultant space between the bar and the arm is taken up at the pivot by inserting a sufficient number of washers over the pivot screw. This supports the arm on a plane level with the table.

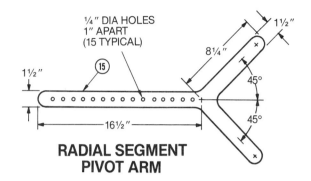

¼" DIA HOLES
1" APART
(15 TYPICAL)

1½"

8¼"

1½"

45°

45°

16½"

RADIAL SEGMENT PIVOT ARM

ITEM	QTY	DESCRIPTION
15	1	¾" × 12" × 24" plywood

Arm ends are tack nailed to the workpiece. Washers stacked over the pivot screw keep arm on a level plane with the table.

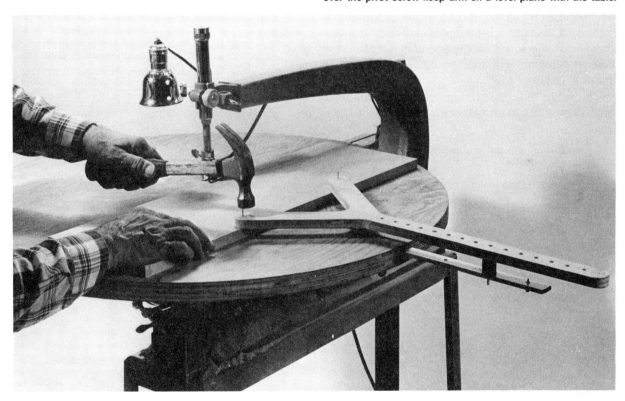

Cutting curved segments the easy way. A change of the pivot location will change the radius of the curve.

5. BAND SAW

Band Saw Extension Table

Better control of the work results in better work and an extension table on your band saw will provide the means to that end. This table is somewhat similar to, though less sophisticated than, the one shown for the scroll saw (page 23). Actually, the basic designs are interchangeable, so both can be adapted to either saw.

The table mounts quickly with four hex-head bolts driven into existing threaded holes in the table sides. Most band saws have these holes for mounting an accessory rip fence. The dimensions shown apply for the Rockwell 14-inch saw. You can alter the dimensions for other saws.

MATERIALS LIST		
ITEM	QTY	DESCRIPTION
1	1	1/2" × 36"-dia. plywood
2	8	1 1/2" # 12 fh wood screw
3	2	2×2 × 32" pine
4	1	2×2 × 14 1/4" pine
5	2	1/4" × 2 1/2" hex-hd bolt
6	2	1/4" flat washer

Note: the actual dimension of 2 × 2 is 1 1/2" × 1 1/2".

Bulky workpieces present no problem when they are supported on this big table.

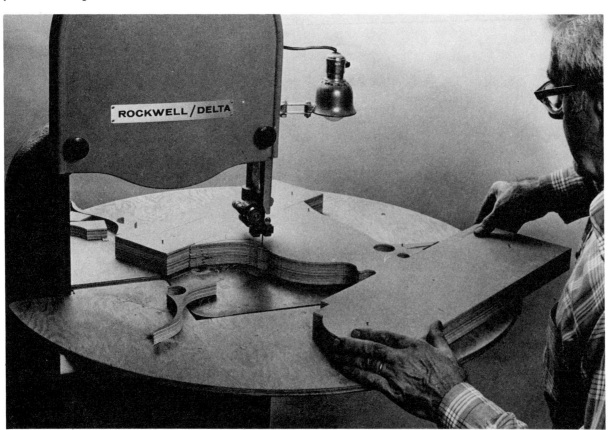

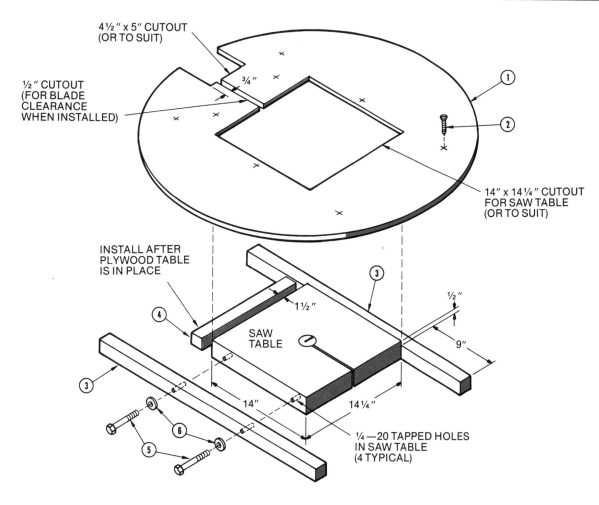

4½" x 5" CUTOUT
(OR TO SUIT)

½" CUTOUT
(FOR BLADE
CLEARANCE
WHEN INSTALLED)

¾"

14" x 14¼" CUTOUT
FOR SAW TABLE
(OR TO SUIT)

INSTALL AFTER
PLYWOOD TABLE
IS IN PLACE

½"

9"

1½"

SAW
TABLE

14"

14¼"

¼—20 TAPPED HOLES
IN SAW TABLE
(4 TYPICAL)

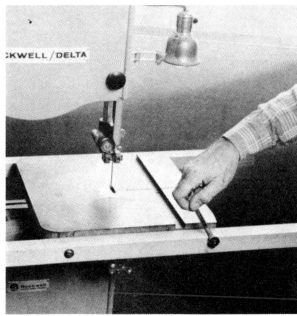

Table supports are attached first so they will be in the correct position when the top is added. Thin washer shims are inserted between the table edges and supports to allow clearance. They are not needed when the table is installed for use.

Screws and glue are used to attach the table to the supports. A 2 × 2 brace (out of view) bridges the blade clearance slot. It is screwed in from below after the table is in place.

Dowel Slicer

Halving or quartering a dowel is a tricky and often frustrating operation, but the task is accurately and safely performed on the band saw with this jig.

A V block with a partial kerf cut at one end is positioned over the blade and clamped in place. A metal vane, the same thickness as the blade's kerf, is mounted on a dowel peg. This is inserted in a hole in the block behind the blade. As the dowel is cut its kerf engages the vane, thus keeping the dowel on an even keel.

Use a piece of thin aluminum stock to make the vane. If necessary, file it to the required thickness, then file a slightly tapered lead edge. To install the vane into the dowel, drill a $1/4$-inch-diameter hole about $3/8$ inch deep into one end of the dowel; then cut the slot to receive the vane. Drill a $1/8$-inch-diameter hole through the vane near the bottom. Use epoxy glue to assemble. The holes in the peg and the vane will fill with the glue and key the two for a good bond. Notice that the vane extends slightly beyond the back of the peg. This keeps it from twisting out of alignment.

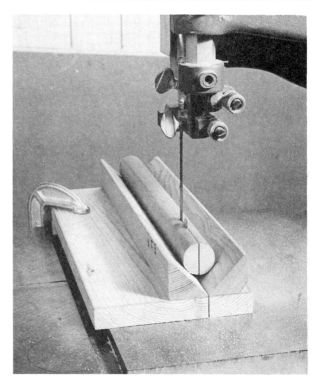

Vane slip-fits the kerf and keeps the round stock straight on course for a perfect cut. The blade guard is raised high for clarity in the photo. Always lower it when cutting.

Pegged vane is inserted in the hole behind the saw blade after the jig has been positioned.

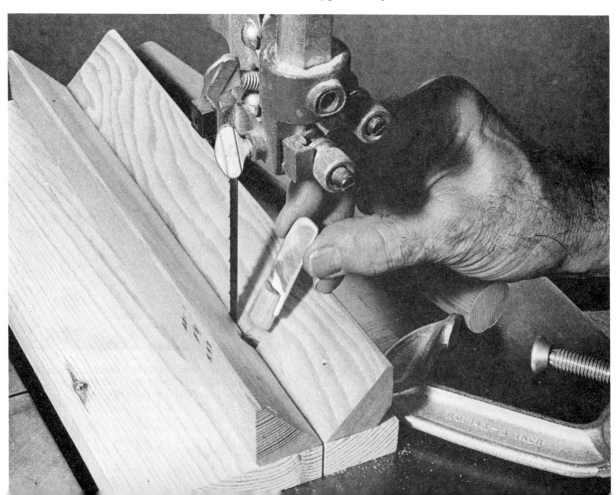

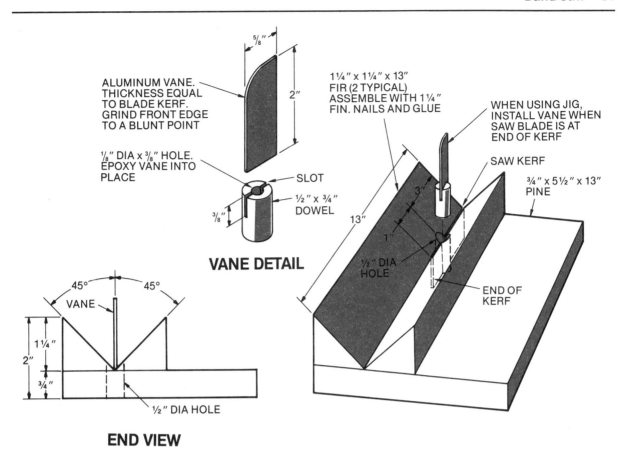

ALUMINUM VANE. THICKNESS EQUAL TO BLADE KERF. GRIND FRONT EDGE TO A BLUNT POINT

5/8"

2"

1/8" DIA x 3/8" HOLE. EPOXY VANE INTO PLACE

SLOT

1/2" x 3/4" DOWEL

3/8"

VANE DETAIL

1 1/4" x 1 1/4" x 13" FIR (2 TYPICAL) ASSEMBLE WITH 1 1/4" FIN. NAILS AND GLUE

WHEN USING JIG, INSTALL VANE WHEN SAW BLADE IS AT END OF KERF

SAW KERF

3/4" x 5 1/2" x 13" PINE

3"

13"

1"

1/2" DIA HOLE

END OF KERF

45°   45°

VANE

1 1/4"

2"

3/4"

1/2" DIA HOLE

END VIEW

Resawing Guide

These vertical supports will not automatically result in a first-rate resawing cut, but they surely will help. Clamp them in place to position the work so the blade centers on the desired cutting line. Feed slowly and continually; manipulate the work as required to keep the blade on the line.

Guides are positioned about 1/4 inch ahead of the blade and spaced apart equal to the thickness of the work. The blade guard is raised for clarity in the photo; *do not* work with the guard raised.

3 1/2"

THIS PIECE REQUIRED ONLY ON INSIDE GUIDE TO AID IN CLAMPING TO BAND SAW TABLE

5/4 HARDWOOD

3 1/2"

1 1/16"

60°

ASSEMBLE WITH 2" FIN. NAILS AND GLUE

PLAN VIEW

6 3/4"

3/4"   1 1/2"   4 1/2"

2 REQUIRED

3"

4 1/2"

1" R

1 1/16"

1 1/2"

FRONT VIEW

Four-Size Radius Jig

When you use this jig to cut corner radii, the edges will be quite smooth and the curves will be perfectly tangent to the straight edges. The open-ended pivoting platform will accept workpieces of unlimited size (within reason, of course).

The jig consists of a base with three side cleats that butt against the saw table to position it automatically and accurately in relation to the blade. A dowel pivot fits in one of four holes that relate to the four position holes in the platform. By matching the appropriate pivot holes, radii of 2, $2^1/2$, 3, and $3^1/2$ inches may be obtained.

To make a cut, the platform is swung against the rear stop block; then it is pivoted until it strikes the front stop. The platform workpiece stop cleats must be repositioned according to the radius position selected.

The jig is sized for the Rockwell 14-inch band saw using a $3/8$-inch blade. The dimensions can be altered to suit other machines.

MATERIALS LIST

ITEM	QTY	DESCRIPTION
1	1	$3/4" \times 11^1/4" \times 11^1/4"$ pine
2	1	$1/2" \times 14" \times 14"$ plywood
3	1	$3/4" \times 1^1/4" \times 14"$ pine
4	1	$3/4" \times 1^1/4" \times 13^1/4"$ pine
5	2	$3/4" \times 3/4" \times 1^3/4"$ pine
6	2	$1/2" \times 1^1/2" \times 8^1/4"$ pine
7	4	1" # 14 pan-hd sheet-metal screw
8	12	$1^1/4"$ finishing nail
9	1	$3/8" \times 1^1/4"$ dowel

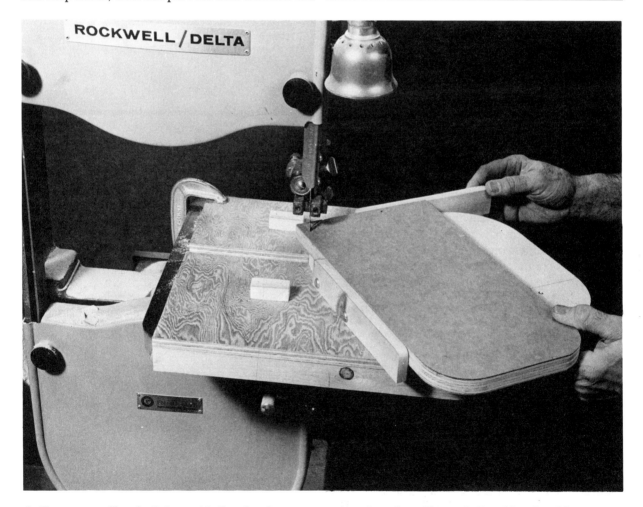

Cutting corner radii perfectly tangent to the edges is an easy task with this jig. Base is custom-fitted to the saw and needs only a clamp for setting up. Lettered dowel positions correspond to radius selections on the platform.

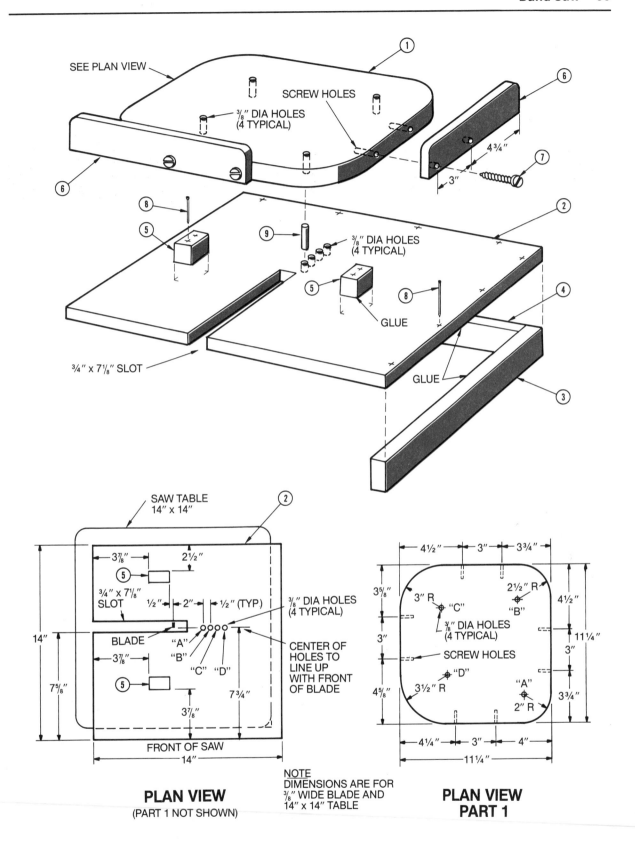

SEE PLAN VIEW

SCREW HOLES

⅜″ DIA HOLES
(4 TYPICAL)

4¾″

3″

⅜″ DIA HOLES
(4 TYPICAL)

GLUE

GLUE

¾″ x 7⅛″ SLOT

PLAN VIEW

SAW TABLE
14″ x 14″

3⅞″

2½″

¾″ x 7⅛″
SLOT

½″

2″

½″ (TYP.)

⅜″ DIA HOLES
(4 TYPICAL)

14″

BLADE

"A"

"B"

"C" "D"

CENTER OF
HOLES TO
LINE UP
WITH FRONT
OF BLADE

3⅞″

7⅝″

3⅞″

7¾″

FRONT OF SAW

14″

PLAN VIEW
(PART 1 NOT SHOWN)

NOTE
DIMENSIONS ARE FOR
⅜″ WIDE BLADE AND
14″ x 14″ TABLE

4½″

3″

3¾″

3⅝″

3″ R

"C"

2½″ R

"B"

4½″

⅜″ DIA HOLES
(4 TYPICAL)

11¼″

3″

3″

SCREW HOLES

4⅝″

3½″ R

"D"

"A"

3¾″

2″ R

4¼″

3″

4″

11¼″

**PLAN VIEW
PART 1**

6. TABLE AND RADIAL-ARM SAW

Feather Board

The feather board, or spring board as it is sometimes called, is a handy device used to apply side pressure to the workpiece in order to keep it continually in contact with the rip fence. It is made with a piece of straight grained wood cut at an angle of about 40 degrees on one end. A series of evenly spaced slots are cut into the end to form slightly flexible fingers.

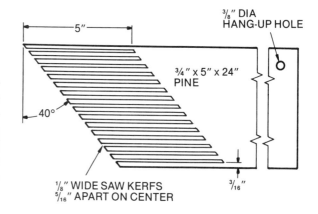

³⁄₈″ DIA
HANG-UP HOLE

5″

³⁄₄″ x 5″ x 24″
PINE

40°

¹⁄₈″ WIDE SAW KERFS
⁵⁄₁₆″ APART ON CENTER

³⁄₁₆″

Feather board is clamped to the saw table so fingers bear against the workpiece with light to moderate pressure, depending upon the size of the workpiece. For through ripping it should be positioned just before the blade and never opposite the blade because this could cause binding and kick-back. However, for a grooving cut such as is being made here, it is permissible to locate the board partially opposite the forward side of the blade.

Tenoning Jig

The tenoning jig is one of the most important accessories for the table saw. If you have not purchased a commercially available jig due to the high cost, you undoubtedly have been relying on makeshift methods for making various joint cuts. Some of the procedures may be unsafe as well as time-consuming; possibly they produce results that are not quite satisfactory.

This homemade jig will enable you to cut mortises, tenons, grooves, bevels, and other precision

Tongue-and-groove and numerous other joint cuts are made safely with speed and precision.

The jig can handle stock up to 12 inches wide; maximum thickness of stock that can be held is 3¼ inches.

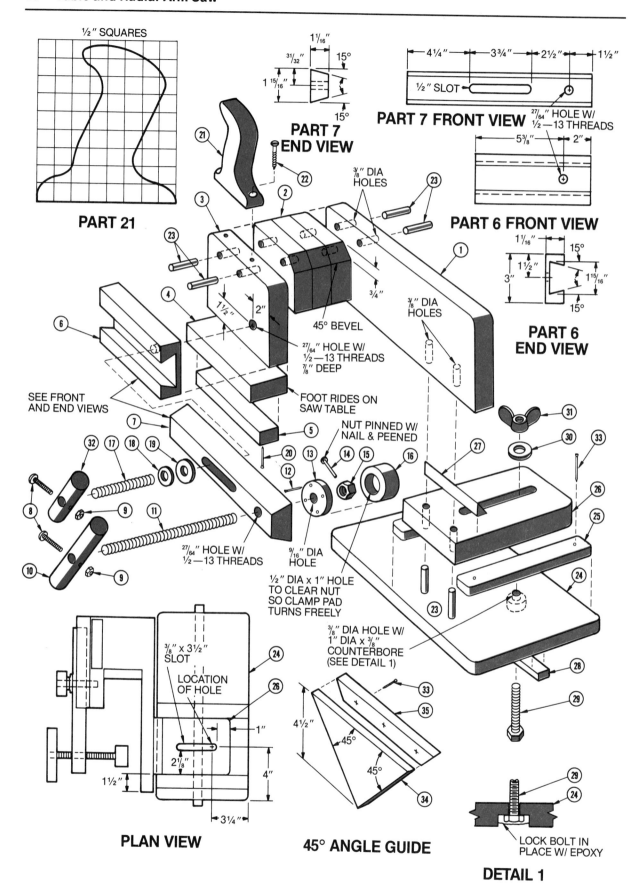

½" SQUARES

PART 21

1 1/16"

31/32"

1 15/16"

15°

15°

**PART 7
END VIEW**

4¼"   3¾"   2½"   1½"

½" SLOT

PART 7 FRONT VIEW

27/64" HOLE W/
½—13 THREADS

5 3/8"   2"

PART 6 FRONT VIEW

1 1/16"

1½"

3"

15°

1 15/16"

15°

**PART 6
END VIEW**

⅜" DIA
HOLES

⅜" DIA
HOLES

¾"

45° BEVEL

2"

1½"

27/64" HOLE W/
½—13 THREADS
⅞" DEEP

FOOT RIDES ON
SAW TABLE

SEE FRONT
AND END VIEWS

NUT PINNED W/
NAIL & PEENED

27/64" HOLE W/
½—13 THREADS

9/16" DIA
HOLE

½" DIA x 1" HOLE
TO CLEAR NUT
SO CLAMP PAD
TURNS FREELY

⅜" DIA HOLE W/
1" DIA x ⅜"
COUNTERBORE
(SEE DETAIL 1)

⅜" x 3½"
SLOT

LOCATION
OF HOLE

1"

2⅛"

4"

1½"

3¼"

PLAN VIEW

4½"

45°

45°

45°

45° ANGLE GUIDE

LOCK BOLT IN
PLACE W/ EPOXY

DETAIL 1

TENONING JIG

MATERIALS LIST

ITEM	QTY	DESCRIPTION
1	1	$1^1/_{16}$″ × $4^3/_4$″ × $14^1/_2$″ ash
2	3	$1^1/_{16}$″ × 3″ × $5^1/_8$″ ash
3	1	$1^1/_{16}$″ × $4^{11}/_{16}$″ × $7^3/_8$″ ash
4	1	$1^1/_{16}$″ × $2^1/_8$″ × $7^3/_8$″ ash
5	1	$3/_4$″ × $3/_4$″ × $7^3/_8$″ ash
6	1	$1^1/_{16}$″ × 3″ × $7^3/_8$″ ash
7	1	$1^1/_{16}$″ × $1^{15}/_{16}$″ × 12″ ash
8	2	8–32 × 1″ rh machine screw
9	2	8–32 hex nut
10	1	$7/_8$″ × 4″ dowel
11	1	$1/_2$–13 × $6^3/_4$″ threaded rod
12	4	$3/_4$″ brad
13	1	$3/_8$″ × $1^3/_4$″-dia. ash
14	1	$1^1/_4$″ nail
15	1	$1/_2$″ hex nut
16	1	$3/_4$″ × $1^3/_4$″-dia. ash
17	1	$1/_2$–13 × $2^3/_4$″ threaded rod
18	1	$1/_2$″ × 1″ flat washer
19	1	$1/_2$″ × $1^1/_4$″ flat washer
20	3	$1^1/_2$″ finishing nail
21	1	$1^1/_{16}$″ × 4″ × $4^1/_2$″ ash
22	2	$1^1/_4$″ # 8 fh wood screw
23	6	$3/_8$″ × 2″ dowel
24	1	$3/_4$″ × 8″ × 15″ plywood
25	2	$3/_8$″ × $1^1/_8$″ × $7^3/_4$″ ash
26	1	$1^1/_6$″ × $4^5/_8$″ × $7^1/_2$″ ash
27	1	1″ × 1″ × $4^5/_8$″ ash
28	1	$3/_8$″ × $3/_4$″ × $15^3/_4$″ ash (or to suit)
29	1	$3/_8$″ × $2^1/_2$″ hex-hd bolt
30	1	$3/_8$″ flat washer
31	1	$3/_8$″ wing nut
32	1	$7/_8$″ × 2″ dowel
33	7	1″ nails
34	1	$1/_2$″ × $4^1/_2$″ × $9^1/_8$″ plywood
35	1	$3/_4$″ × 1″ × $9^1/_8$″ pine

Note: $1^1/_{16}$″ is the usual thickness of $5/_4$ stock.

This sturdy homemade jig is as efficient as any commercially made counterpart.

cuts quickly and safely. The dimensions shown apply to a 10-inch Rockwell Unisaw. This has a $3/_8$-by-$3/_4$-inch miter gauge groove $4^1/_2$ inches from the left side of the blade. The size and position of the miter groove bar will likely have to be altered for other saws. Also, if you plan to make the jig for a 12-inch saw, it will be necessary to increase the blade clearance under the "bridge."

The slots in the clamp bar and the sliding platform are made by using a brad point or spur drill bit to bore a series of slightly overlapping holes. The webs are then cleaned out with a chisel.

Chisel is used to clean out the webs between the overlapping holes in clamp bar.

The dovetail slot for the clamp bar is made with repeated saw cuts. The blade is tilted and several cuts are made from each side; the blade is then zeroed and the height adjusted to clear out the waste (the blade will project more when it is brought back to vertical).

To function properly it is essential that the work-bearing wall be perfectly perpendicular and the carriage must track at a precise right angle to the blade. Check these critical relationships by dry-assembling the parts with screws before final gluing. To obtain accuracy in positioning the sliding carriage guide strips, raise the blade to maximum height, then butt the vertical wall up to it.

It is important to bore the hole to the proper size to tap lasting threads in hardwood. Use a $^{27}/_{64}$-inch drill bit and cut the threads in the clamp assembly with a $^{1}/_{2}$–13 NC tap. Assemble the two parts of the clamp pad in place over a nut that is pin locked on the threaded rod. The holes in both sections must be slightly oversized to allow the clamp pad to rotate freely.

Finish the wood with a clear satin polyurethane coating, then buff with extra-fine steel wool. Apply paste wax, well rubbed in, to the sliding surfaces.

Repeated bevel and straight kerf cuts are made to form the dovetail. The scrap block taped to the workpiece prevents it from tipping as it passes through.

Work-bearing wall is butted against the raised blade to align the platform guide strips accurately.

Coarse threads cut in hardwood will hold up very well.

When using a brad point bit it is necessary to drill the counter-bore hole first, then the screw body hole and finally the screw shank hole.

Precision in assembly is important. Parts are trued with a square and clamped together to drill the screw pilot holes.

Clamp pad is assembled in place over the pin-locked nut. The parts are held together with brads after glue is applied.

A 45-degree-angle guide is an optional accessory.

Guide is used to position the work in a true 45-degree angle. When clamped in place it provides a more positive method of aligning the work than is possible by using a hand-held T bevel or combination square.

Blade Projection Gauge

This gauge will give you positive direct readings of your table-saw-blade height adjustment. When the bottom of the floating dowel is placed over the high point of the blade it rides up or down with it. Sighting on the fine crossline permits adjustments in increments of $1/32$ inch by reading between the rule's $1/16$-inch graduations.

Begin construction with the top block. Drill a $7/8$-inch-diameter hole. Then sand a dowel, if necessary, to slip-fit the hole. Make a stopped-kerf cut in the vertical member. Position and attach the top block so the hole sets back $1/8$ inch past the inside edge of the vertical. Glue the bottom of the vertical into the notch in the base.

Form a flat on the dowel by planing, sanding, or sawing so it will pass through the hole in the top block. Drill a blind hole centered on the flat of the

dowel $7/8$ inch from the bottom to receive a 2-inch common nail. Cut the nail to length so its head just about touches the upright when it is inserted through the kerf and into the dowel. This serves as a retainer, keeping the dowel in line as it moves up and down.

Make the crossline by scribing a line on the surface of a piece of $1/8$-inch-thick Plexiglas or similar heat-formable acrylic plastic. The fine line can be made with a sharp awl or a plastics-scribing tool.

MATERIALS LIST		
ITEM	QTY	DESCRIPTION
1	1	$3/4'' \times 2'' \times 2^{1}/4''$ pine
2	1	$3/4'' \times 2'' \times 6''$ pine
3	1	$3/4'' \times 2^{5}/8'' \times 6''$ pine
4	1	$7/8'' \times 7''$ dowel
5	1	$1/8'' \times 1^{1}/8'' \times 3^{1}/4''$ Plexiglas (or equiv.)
6	2	$3/4''$ # 6 rh wood screw
7	1	2'' common nail (cut to $1^{1}/4''$)
8	4	$1^{1}/2''$ finishing nail
9	1	flat plastic rule with $1/16''$ graduations

Precise blade height adjustments are easy with this gauge.

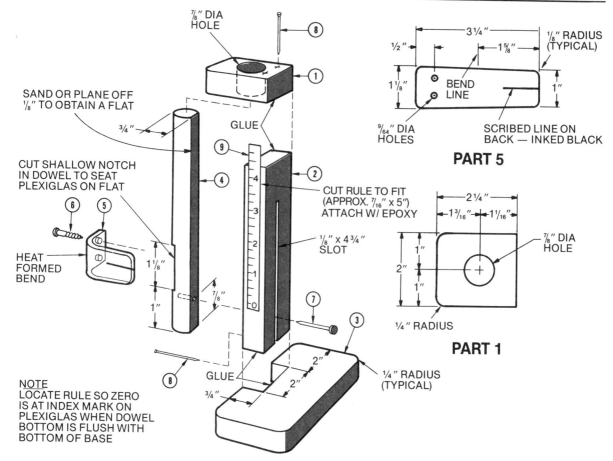

PART 5

PART 1

SAND OR PLANE OFF
⅛" TO OBTAIN A FLAT

CUT SHALLOW NOTCH
IN DOWEL TO SEAT
PLEXIGLAS ON FLAT

HEAT
FORMED
BEND

GLUE

CUT RULE TO FIT
(APPROX. ⁷⁄₁₆" x 5")
ATTACH W/ EPOXY

⅛" x 4¾"
SLOT

GLUE

NOTE
LOCATE RULE SO ZERO
IS AT INDEX MARK ON
PLEXIGLAS WHEN DOWEL
BOTTOM IS FLUSH WITH
BOTTOM OF BASE

⅞" DIA
HOLE

⅛" RADIUS
(TYPICAL)

BEND
LINE

⁹⁄₆₄" DIA
HOLES

SCRIBED LINE ON
BACK — INKED BLACK

¼" RADIUS

⅞" DIA
HOLE

¼" RADIUS
(TYPICAL)

Bent end of plastic is held in position during cooling to prevent it from springing back.

Masking tape is used to hold the rule in position while the epoxy adhesive sets.

Use a small hair dryer or a flameless heat gun to form the right-angle bend in the plastic. Tape the plastic onto a piece of wood with a rounded corner, allowing the end to overhang about an inch. The surface with the scribed line should face down on the wood. Apply heat until the plastic becomes limp and drapes over the corner. When the bend has formed, remove the heat and hold a piece of wood against the end until the piece cools.

Ink over the crossline with a black felt-tip pen, then quickly wipe over the surface with a tight pad of cloth dampened with alcohol so the ink remains in the recessed crossline only.

Cut a strip from a thin, flat, plastic rule and roughen the back slightly by sanding. Position the rule so it reads zero when the dowel is flush on the table surface, then attach it with quick-setting epoxy glue.

Hollow Molding Mitering Jig

As described earlier, but worth mentioning again, crown, bed, and cove moldings are designed for installation in a sloped position. This necessitates compound-angle cuts in order to achieve 45-degree mitered corner joints and can frustrate many who work by trial and error; it results in the waste of time and material regardless of whether the cuts are attempted by hand or on stationary power saws. You need fumble no longer with the problem of the complex, compound-angle cut if you make one of the jigs shown here. One is for use on the table saw; the other is for use with the radial-arm saw. Both are designed to support the molding during the cut in the same sloped position in which it will be installed. Thus, a normal 45-degree miter cut automatically produces the correct compound-angled edge.

Both jigs have backboards that are sized for the capacity of 10-inch saws; therefore the size of the molding that can be accommodated is limited to the inside height of the backboard relative to the molding in its leaning position. In either jig the inside support board is not permanently mounted; instead, it is tack nailed into place to permit shifting locations to suit moldings of varying sizes. A sheet of 220-grit abrasive paper is fixed to the top surface of the base with rubber cement to prevent the molding from creeping during cutting.

Left and right miters are cut on the table saw by swinging the miter gauge left or right as required.

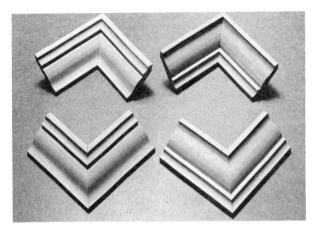

Samples of the various mitered corner treatments produced with the jigs shown here.

Easily constructed mitering jigs for radial-arm and table saws.

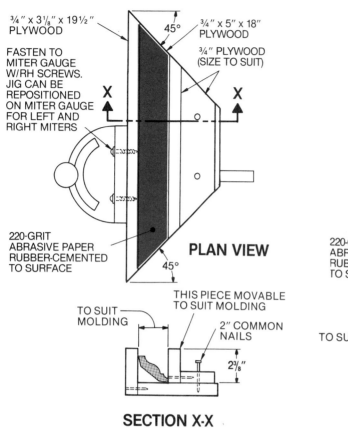

¾" x 3⅛" x 19½" PLYWOOD

FASTEN TO MITER GAUGE W/RH SCREWS. JIG CAN BE REPOSITIONED ON MITER GAUGE FOR LEFT AND RIGHT MITERS

45°

¾" x 5" x 18" PLYWOOD

¾" PLYWOOD (SIZE TO SUIT)

X

X

220-GRIT ABRASIVE PAPER RUBBER-CEMENTED TO SURFACE

45°

PLAN VIEW

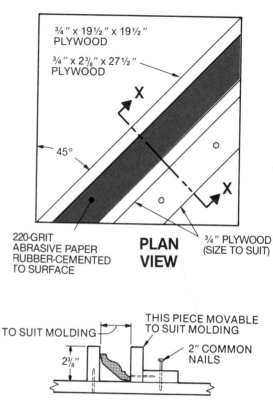

¾" x 19½" x 19½" PLYWOOD

¾" x 2⅜" x 27½" PLYWOOD

X

45°

X

220-GRIT ABRASIVE PAPER RUBBER-CEMENTED TO SURFACE

PLAN VIEW

¾" PLYWOOD (SIZE TO SUIT)

TO SUIT MOLDING

THIS PIECE MOVABLE TO SUIT MOLDING

2" COMMON NAILS

2⅜"

SECTION X-X

TO SUIT MOLDING

THIS PIECE MOVABLE TO SUIT MOLDING

2" COMMON NAILS

2⅜"

SECTION X-X

Fine abrasive paper is rubber-cemented to the base to prevent the molding from creeping during cutting (due to side thrust of blade against the angled work).

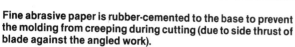

Table saw jig in use. Cutoff mark on the molding is easily aligned to the edge of the jig. This reduces chance of error.

Slanted molding is held in position by tack nailing the inside support to base.

Two sets of screw pilot holes are needed for attaching the jig to the miter gauge because it must be shifted into position to bring its left or right mitered edge in line with the saw blade.

Left and right miters are cut on the radial-arm saw by alternately shifting the jig to either side of the blade with a quarter-turn rotation.

An important feature of both jigs is that cuts to exact lengths can easily be made because the cutting marks on the work need only to be visually aligned to the mitered edge of the jig.

This is the setup for cutting a right miter on the radial-arm saw.

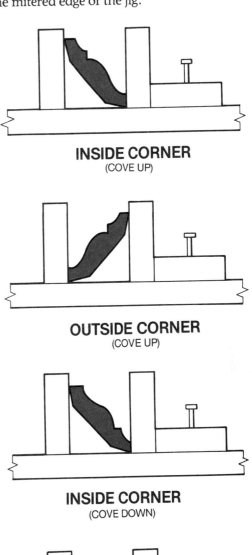

INSIDE CORNER
(COVE UP)

OUTSIDE CORNER
(COVE UP)

INSIDE CORNER
(COVE DOWN)

OUTSIDE CORNER
(COVE DOWN)

Jig is shifted to the right side of the blade and rotated a quarter turn to cut a left miter.

Adjustable Hold-Down

threads. But in this application, where there is little surrounding wood, it is necessary to tap threads.

A sturdy hold-down is particularly important for dadoeing and grooving operations to ensure a uniform depth of cut. It is also helpful in preventing chattering when sawing thin or small pieces of stock. And most important, it allows you to keep your fingers at a safe distance from the blade.

The unit attaches to the fence with two flat head screws driven through countersunk holes in the rip fence. The base can be left on the fence permanently. The sliding hold-down assembly is easily removed when it is not needed by withdrawing the wing nut. When installing the base, be sure to position it on the fence so the pressure foot on the sliding assembly will locate just before the blade.

The dowel post is faced with a steel strip to provide a good bearing surface for the locking thumbscrew. The thumbscrew threads into a threaded steel insert (#15 on plan) installed in the end of the sliding bar. Normally these inserts can be turned into snug holes in which they form their own

MATERIALS LIST

ITEM	QTY	DESCRIPTION
1	1	$^{13}/_{16}$" × $3^1/_8$" × 8" oak
2	1	$^{13}/_{16}$" × $1^1/_2$" × 5" oak
3	1	$^9/_{16}$" × $^{13}/_{16}$" × 5" oak
4	1	$^{13}/_{16}$" × $3^1/_8$" × 5" oak
5	2	$^3/_8$" × $^7/_8$" × $2^7/_8$" oak
6	1	$^{13}/_{16}$" × $2^1/_8$" × $8^{15}/_{16}$" oak
7	1	$^9/_{16}$" × $^{13}/_{16}$" × $2^1/_8$" oak
8	1	1" × $4^3/_4$" hardwood dowel
9	1	$^1/_{16}$" × $^1/_2$" × $4^5/_8$" steel strip
10	1	$^1/_8$" × 1" × $5^1/_2$" mild steel (hot-rolled)
11	1	$^1/_4$" × $1^1/_2$" lag screw
12	1	$^1/_4$" lock washer
13	2	$^1/_2$" # 6 fh wood screw
14	1	$^1/_4$–20 × 1" thumbscrew
15	1	$^1/_4$–20 steel threaded insert
16	1	$^3/_8$" wing nut
17	1	$1^1/_2$" fender washer
18	1	$^3/_8$" × 3" carriage bolt
19	2	2" # 10 fh wood screw
20	2	$1^1/_2$" # 12 fh wood screw

The foot applies downward pressure where it does most good —close to the blade. This is particularly important when making groove cuts for dado joints.

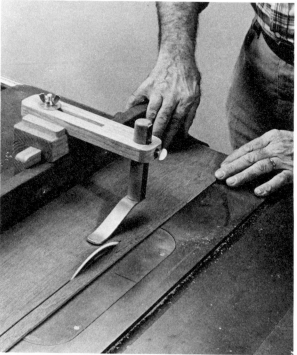

The foot can slide outward up to 3¾ inches from the fence. Thickness of stock capacity is 2¾ inches. Push stick will be used to follow through the feed of this cut.

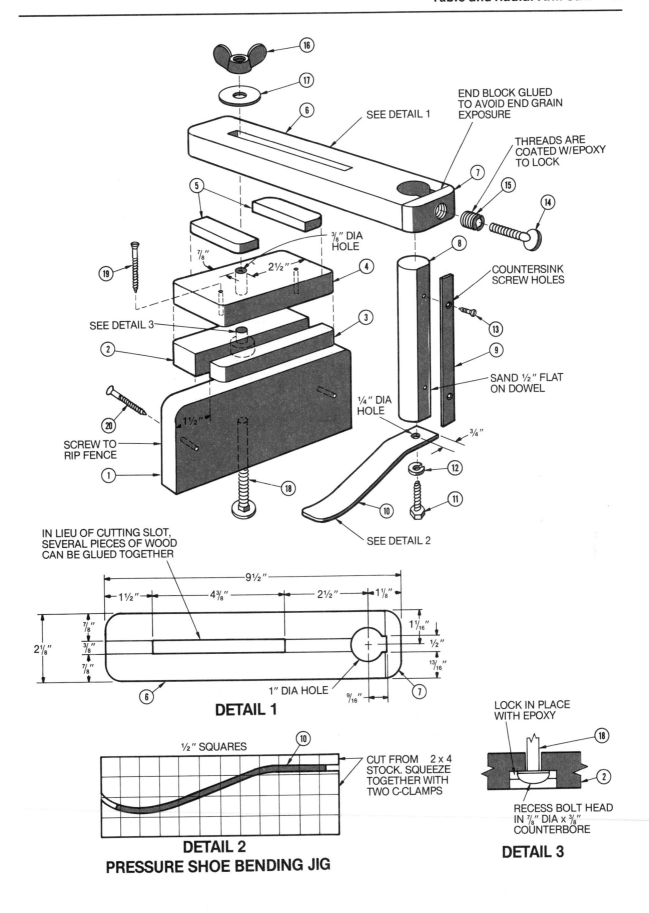

END BLOCK GLUED
TO AVOID END GRAIN
EXPOSURE

SEE DETAIL 1

THREADS ARE
COATED W/EPOXY
TO LOCK

³⁄₈″ DIA
HOLE

⁷⁄₈″

2½″

COUNTERSINK
SCREW HOLES

SEE DETAIL 3

¼″ DIA
HOLE

SAND ½″ FLAT
ON DOWEL

1½″

¾″

SCREW TO
RIP FENCE

SEE DETAIL 2

IN LIEU OF CUTTING SLOT,
SEVERAL PIECES OF WOOD
CAN BE GLUED TOGETHER

9½″

1½″   4³⁄₈″   2½″   1¹⁄₈″

1¹⁄₁₆″

2¹⁄₈″

⁷⁄₈″
³⁄₈″
⁷⁄₈″

½″

¹³⁄₁₆″

1″ DIA HOLE

⁹⁄₁₆″

DETAIL 1

½″ SQUARES

CUT FROM   2 x 4
STOCK. SQUEEZE
TOGETHER WITH
TWO C-CLAMPS

DETAIL 2
PRESSURE SHOE BENDING JIG

LOCK IN PLACE
WITH EPOXY

RECESS BOLT HEAD
IN ⁷⁄₈″ DIA x ³⁄₈″
COUNTERBORE

DETAIL 3

To lock the insert firmly in place, file a small notch across its threads and apply a drop of epoxy glue. This will key it and prevent it from ever turning loose. The handy inserts are available from woodworking supply outlets.

A strip of hot-rolled mild steel is used to make the pressure foot. It is bent to shape with C clamps and the forming jig shown in the drawing.

Steel strip is bent between forming blocks in this manner. Curve in the block is slightly exaggerated to allow for springback when the piece is released.

Hole for the threaded insert is tapped to obtain a stronger hold. Notice how the weaker end grain is avoided by the addition of a side grain end strip.

Quick-Lock Stop Block

This self-locking rip fence stop block eliminates the need for fussing with a clamp. It is quickly locked and released by the action of the lever handle. As the handle is brought down, its tapered sides squeeze the jaws firmly against the fence. The pressure is relieved when the handle is lifted. The dimensions shown are for a fence 1 inch thick. Alter them to suit if your fence thickness differs.

Handle is pushed down to lock the block.

MATERIALS LIST		
ITEM	QTY	DESCRIPTION
1	2	3/4" × 2 3/4" × 4" pine
2	4	1/4" × 3/4" × 2 3/8" pine
3	4	1/4" × 3/4" × 7/8" pine
4	1	1" × 1 3/8" × 5 3/8" pine
5	2	1/16" × 1 1/8" × 1 3/8" hardwood
6	1	1/4" × 3" rh machine screw
7	2	1/4" flat washer
8	3	1/4" hex nut
9	2	1/4" × 3" hex-hd bolts
10	2	1/16" × 5/8" × 2 1/2" mending plates

Using the stop for partial rip cuts.

Stop is useful for sizing repeat cutoffs. It should be located so the work clears it before contacting the blade.

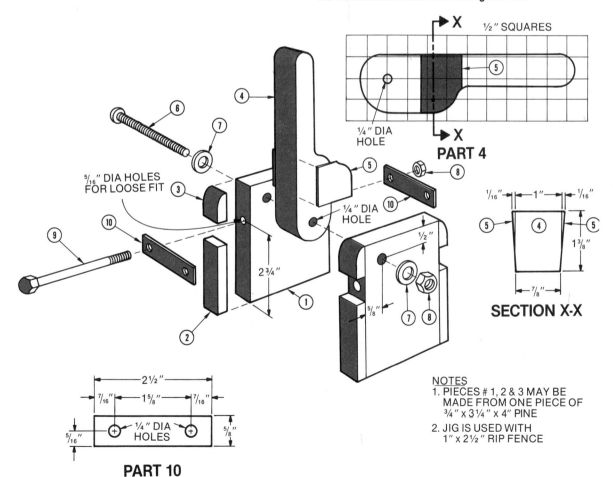

½″ SQUARES

¼″ DIA HOLE

PART 4

⑥

⑦

⑤

⑧

④

⁵⁄₁₆″ DIA HOLES FOR LOOSE FIT ③

⑩

¼″ DIA HOLE

⑨

⑩

2¾″

½″

⑦ ⑧

⁵⁄₈″

②

①

¹⁄₁₆″ 1″ ¹⁄₁₆″

⑤ ④ ⑤

1³⁄₈″

⁷⁄₈″

SECTION X-X

2½″

⁷⁄₁₆″ 1⁵⁄₈″ ⁷⁄₁₆″

¼″ DIA HOLES

⁵⁄₁₆″

⁵⁄₈″

PART 10

NOTES
1. PIECES # 1, 2 & 3 MAY BE MADE FROM ONE PIECE OF ¾″ x 3¼″ x 4″ PINE
2. JIG IS USED WITH 1″ x 2½″ RIP FENCE

Mitering Jig

The sliding platform mitering jig is not a new idea, but this one has a new twist—a unique method for clamping the work to prevent creeping during the cutting action. An adjustable bar and a flexible wedge combine for quick setup and positive holding.

Use plywood for the platform and solid hardwood for the other parts. The shallow recesses on the platform are cut using a straight bit with a router. To save time and effort, the slide bar slot is pre-formed by gluing spacer blocks between the ends of two pieces of stock. A similar technique, using a single tapered block, is used to make the wedge. It should be noted that the wedge has a front and a back strip. The front strip is a bit narrower in order to give it flex. It faces the bar in use.

Attach the miter slot guide strips to the bottom of the platform before installing the angled backstops.

Tack nail one backstop temporarily into place and cut two pieces of stock for a test. If the butted miters check out to form a true right angle, advance the cut into the backstop to miter its end. Remove the first stop and repeat the procedure with the other. Make a final test, then glue and nail the stops permanently.

MATERIALS LIST		
ITEM	QTY	DESCRIPTION
1	1	3/4" × 153/4" × 231/4" plywood
2	2	13/16" × 21/2" × 153/4" hardwood
3	2	13/16" × 17/8" × 8" hardwood
4	2	3/8" × 3/4" × 20" hardwood
5	2	5/16" × 2" carriage bolt
6	2	5/16" flat washer
7	2	5/16" wing nut
8	4	11/4" finishing nail
9	8	1" finishing nail

Bar and flexible wedge hold the work firmly in place to effectively prevent creep.

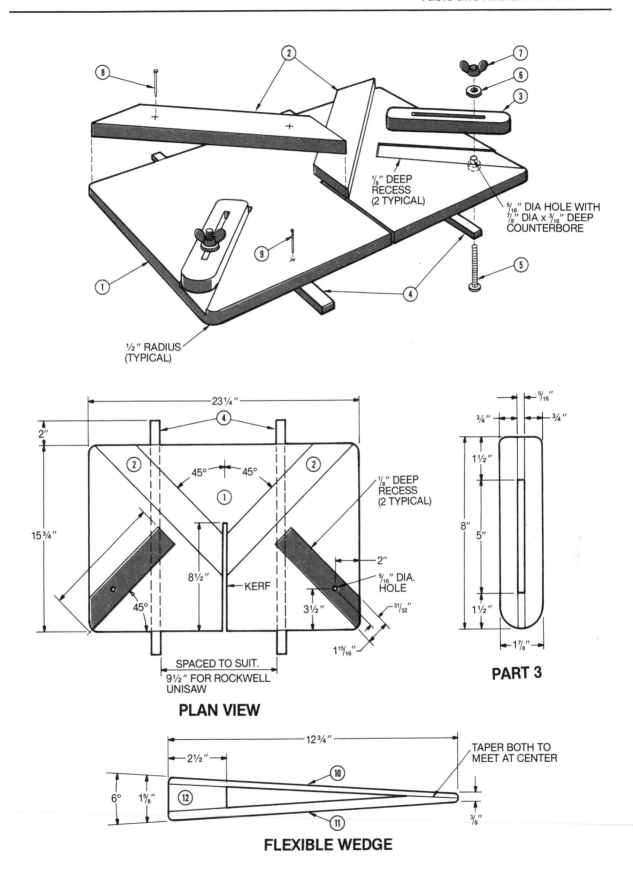

½″ RADIUS
(TYPICAL)

⅛″ DEEP
RECESS
(2 TYPICAL)

5/16″ DIA HOLE WITH
7/8″ DIA x 3/16″ DEEP
COUNTERBORE

23¼″

2″

15¾″

45°   45°

45°

⅛″ DEEP
RECESS
(2 TYPICAL)

2″

5/16″ DIA.
HOLE

8½″

KERF

3½″

31/32″

1 15/16″

SPACED TO SUIT.
9½″ FOR ROCKWELL
UNISAW

PLAN VIEW

5/16″

¾″   ¾″

1½″

8″   5″

1½″

1⅞″

PART 3

12¾″

TAPER BOTH TO
MEET AT CENTER

2½″

6°   1⅝″

⅜″

FLEXIBLE WEDGE

Slide bar is butted against the wedge, then locked in place. Wedge is then pushed slightly for a jam fit.

Backstop is temporarily tack nailed. Its end is not cut off until test miters have been cut.

Tack nail registration holes permit removal and exact realignment after each backboard has been tested and mitered. When all checks out, glue is applied.

Circular Work Edging Jig

The shaping of circular edges is generally a task accomplished with a router, but there may be times when the table saw will be the better or only choice for making the cut. If you lack a router bit of a particular shape, the table saw molding cutterheads, the dado head, or the regular saw blade may solve the problem with the aid of one of the jigs shown here. Two of them can be used when it is feasible to bore a pivoting hole through the center of the work; the other is used when a through hole is not desired.

Work is rotated opposite the rotation of the blade. Deep molding cuts should be made in small bites by elevating the blade in small increments after each full rotation of the work. Work is lifted clear of the blade before each succeeding pass.

The pivoted backboard arcs to bring pivoted work into contact with the blade. Power is turned on *before* contact is made.

Cutting a groove with a dado head. This jig consists merely of a high auxiliary fence clamped to the rip fence; a nail serves as a pivot. For this operation a helper should turn on the power while the work is held firmly in place.

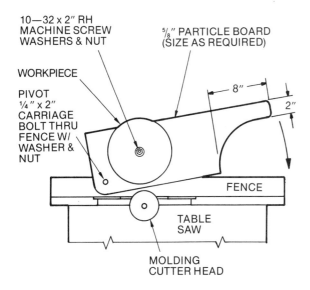

10—32 x 2″ RH MACHINE SCREW WASHERS & NUT

⁵⁄₈″ PARTICLE BOARD (SIZE AS REQUIRED)

WORKPIECE

PIVOT ¼″ x 2″ CARRIAGE BOLT THRU FENCE W/ WASHER & NUT

8″

2″

FENCE

TABLE SAW

MOLDING CUTTER HEAD

NOTE
CENTER OF WORK TO LINE UP WITH CENTER OF CUTTER WHEN JIG IS IN HORIZONTAL POSITION

Cutting a rabbet on the edge of a circular disc. V cut in the backboard provides clearance for gripping and spinning the work. Front board has a circular cutout with the same radius as that of the workpiece. The saw blade makes its own recess in the lower section as it is elevated.

Spherical Dishing Jig

The table saw can be used to cut a perfectly spherical depression in a circular workpiece with this jig —a piece of plywood with a V cut out.

To make the jig, use a piece of ³/₈-inch plywood long enough to span the table top and permit clamping at both ends. Cut a right-angle notch set at 45 degrees to the edge.

Position the notch so the work disc centers directly over the center of the blade when its edge is butted against the notch. Clamp the jig in place.

To make the cut, depress the blade to just below the table surface. Then, holding the work firmly with one hand, turn on the power and elevate the blade until it contacts the work slightly. Rotate the work a full revolution, then elevate the blade again and repeat the procedure. Continue making the cuts, elevating the blade in small increments of no more than about ¹/₈ inch at a time, until the desired depth of cut is reached. Never make a deep cut in a single pass; this could be dangerous. On most saws a quarter turn of the elevating crank will raise the blade about ¹/₈ inch, but check yours out in advance. An important note: make constant checks of the depth of the cut as the operation progresses. You need to be careful not to cut through the top to avoid meeting the blade with your hand.

Slow and easy does it. The disc must always bear against the jig while it is rotated.

The perfectly shaped depression will have the same radius as that of the saw blade.

Blade Jointing Aids

When a circular blade needs jointing the job is done with the blade in place on the saw. Lower the arbor until the blade is below, but not touching, an oilstone supported on two wood blocks. Raise the arbor while firmly holding the stone down until the blade teeth just touch the stone.

Slowly rotate the blade backwards by hand, until all the teeth show that they have been touched by the stone. On a radial-arm saw the stone is rested on the table and the blade is brought down to it. For a portable saw you simply rest the stone on the base. In all cases, *never* perform this operation with the blade under power.

Bridge the stone on two blocks and rotate the blade backwards *by hand* to grind the teeth points lightly. If the blade drive belt is out of reach, you will need an assistant.

Raker teeth of combination blades must be jointed separately. The file is placed on a block and slid forward to obtain a square, true cut. Wedges sandwiched over blade hold it in place while filing.

The raker teeth of a combination blade must be filed down individually. Block the blade so it remains stationary and file squarely across the top to reduce its height by about ¹/₆₄ inch relative to the regular teeth. Support the file on a small block of wood and slide the block along the table to ensure accuracy.

Extension Table

Ideally, the radial-arm saw should be adjoined by fair-sized built-in tables on each side to facilitate handling long workpieces. But space limitations often prevent this luxury.

You can make do very nicely, however, with this knockdown extension table. It can be set up in seconds when needed, and it folds flat for convenient storage.

Exercise care in sizing the coupling blocks so they will mate without excessive play. Also, size the leg length so it will support the table top on the same plane as the saw table. The fence is attached with the table in place to ensure alignment with the saw fence.

Coupling blocks are assembled with screws and glue. Two nails with the points cut off serve as pins to join the sections.

This lightweight extension table takes the strain out of handling large work; it also promotes accuracy. A second table can be made for the other side of the saw.

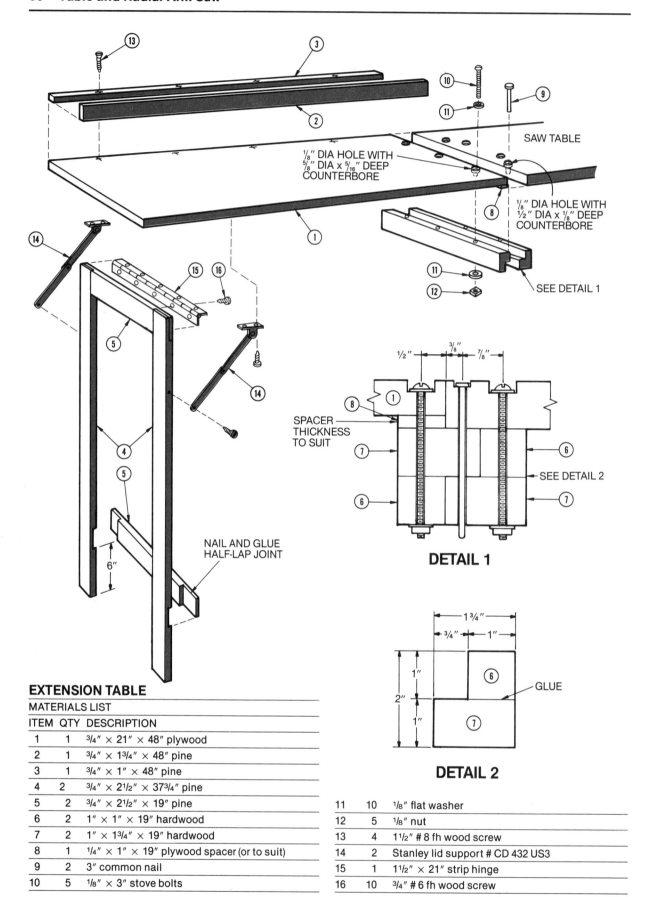

¹⁄₈″ DIA HOLE WITH
⁵⁄₈″ DIA x ⁵⁄₁₆″ DEEP
COUNTERBORE

SAW TABLE

¹⁄₈″ DIA HOLE WITH
¹⁄₂″ DIA x ¹⁄₈″ DEEP
COUNTERBORE

SEE DETAIL 1

NAIL AND GLUE
HALF-LAP JOINT

6″

SPACER
THICKNESS
TO SUIT

SEE DETAIL 2

¹⁄₂″   ³⁄₈″   ⁷⁄₈″

DETAIL 1

1³⁄₄″
³⁄₄″   1″
1″
2″
1″
GLUE

DETAIL 2

EXTENSION TABLE

MATERIALS LIST

ITEM	QTY	DESCRIPTION
1	1	³⁄₄″ × 21″ × 48″ plywood
2	1	³⁄₄″ × 1³⁄₄″ × 48″ pine
3	1	³⁄₄″ × 1″ × 48″ pine
4	2	³⁄₄″ × 2¹⁄₂″ × 37³⁄₄″ pine
5	2	³⁄₄″ × 2¹⁄₂″ × 19″ pine
6	2	1″ × 1″ × 19″ hardwood
7	2	1″ × 1³⁄₄″ × 19″ hardwood
8	1	¹⁄₄″ × 1″ × 19″ plywood spacer (or to suit)
9	2	3″ common nail
10	5	¹⁄₈″ × 3″ stove bolts
11	10	¹⁄₈″ flat washer
12	5	¹⁄₈″ nut
13	4	1¹⁄₂″ # 8 fh wood screw
14	2	Stanley lid support # CD 432 US3
15	1	1¹⁄₂″ × 21″ strip hinge
16	10	³⁄₄″ # 6 fh wood screw

Grooving Jig for Miters

In order to obtain spline grooves in the faces of miters that are in perfect alignment, the same surfaces of the work must bear against the fence. This novel jig is reversible in order to meet the requirement. When one set of miters have been grooved, flip the jig over to permit grooving the faces of the miters of the opposing angle with the same surfaces of the work against the fence. The plan should be adjusted to fit the desired miter angle, and the width and thickness of the work.

Slot in the jig is sized to hold the workpiece snug against the fence as it passes through.

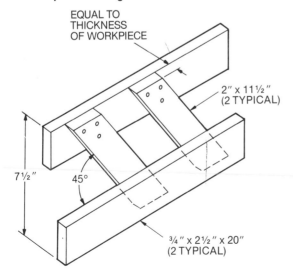

EQUAL TO
THICKNESS
OF WORKPIECE

2" x 11½"
(2 TYPICAL)

7½"   45°

¾" x 2½" x 20"
(2 TYPICAL)

Blade Filling Vise

Working with a dull saw blade puts a strain on the motor, results in poor work, and can be dangerous because it induces kick-back.

Sharpening is relatively easy, but a saw blade cannot be properly filed unless you have a suitable vise that will grip the blade close to the teeth to prevent flexing and chattering. Make this one with an adjustable shaft to handle blades up to 10 inches in diameter.

Thin plywood shoulders on the jaw tops bear against the blade to secure it firmly. Bolt shaft can be positioned for blades of varying diameters.

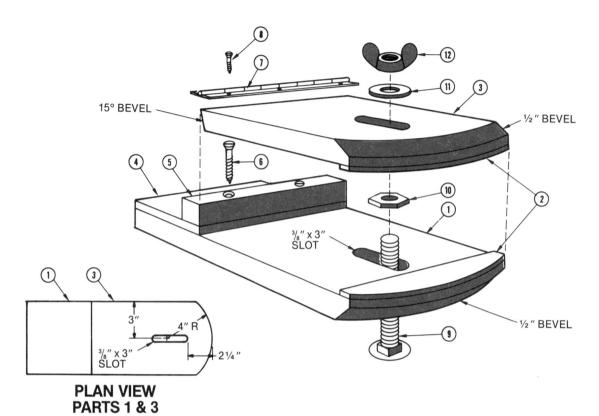

15° BEVEL

½" BEVEL

½" BEVEL

½" BEVEL

$\frac{3}{8}$" x 3"
SLOT

**PLAN VIEW
PARTS 1 & 3**

3"

4" R

$\frac{3}{8}$" x 3"
SLOT

2¼"

BLADE FILING VISE

MATERIALS LIST

ITEM	QTY	DESCRIPTION
1	1	$\frac{3}{4}$" × 6" × 16" plywood
2	2	$\frac{1}{8}$" × 1$\frac{1}{2}$" × 6" hardboard
3	1	$\frac{3}{4}$" × 6" × 10$\frac{1}{2}$" plywood
4	1	$\frac{1}{4}$" × 5$\frac{1}{2}$" × 6" plywood
5	1	$\frac{3}{4}$" × 1$\frac{1}{2}$" × 6" plywood
6	2	1$\frac{1}{2}$" # 8 fh wood screw
7	1	1$\frac{1}{2}$" × 6" strip hinge
8	6	$\frac{3}{4}$" # 6 fh wood screw
9	1	$\frac{3}{8}$" × 2$\frac{1}{2}$" carriage bolt
10	1	$\frac{3}{8}$" hex nut (cut to $\frac{1}{8}$" thickness)
11	1	$\frac{3}{8}$" flat washer
12	1	$\frac{3}{8}$" wing nut

Do not allow blades to become dull; an occasional touch-up filing will do wonders.

Flip-Out Stop

Gauging the length of multiple cutoffs of small pieces of stock with a clamped fixed stop on the radial-arm saw is a common but poor practice. The tiny pieces invariably get jammed between the blade and the stop and become damaged. In the process the pieces are frequently kicked back toward the operator.

The problem can be avoided with the use of this hinged stop. The stop is temporarily flipped back out of the way after the work has been butted up against it for sizing. The hinge used for this application must be of the type with a snug-fitting pin so the block will not wobble.

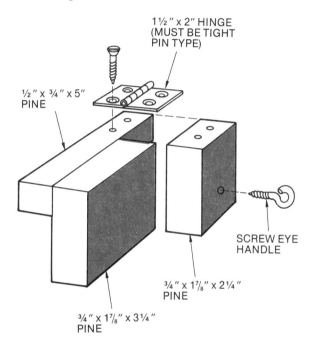

1½" x 2" HINGE
(MUST BE TIGHT
PIN TYPE)

½" x ¾" x 5"
PINE

SCREW EYE
HANDLE

¾" x 1⅞" x 2¼"
PINE

¾" x 1⅞" x 3¼"
PINE

Block is flipped up out of the way while you make the cut.

Workpiece is butted against the stop to gauge the cutoff.

7. PORTABLE DRILL

Angle Jig

You can obtain drill-press precision in drilling angled holes by mounting your portable drill in this jig which utilizes the Portalign drill guide. The tilting base adjusts for drilling holes at any angle up to 35 degrees off perpendicular. Direct reading of the protractor scale will enable you to set the required angle exactly and with ease.

Make the jig with pine or any wood of your choice. The only requirement is that the wood be warp-free. Select a strip hinge with a snug-fitting pin; a hinge with excessive play will affect accuracy. It is important to attach the hinge carefully to avoid a gap between the base and the sub-base.

Special care must also be exercised in positioning

This jig, used in conjunction with a Portalign drill guide, enables you to do precision-angle drilling with ease.

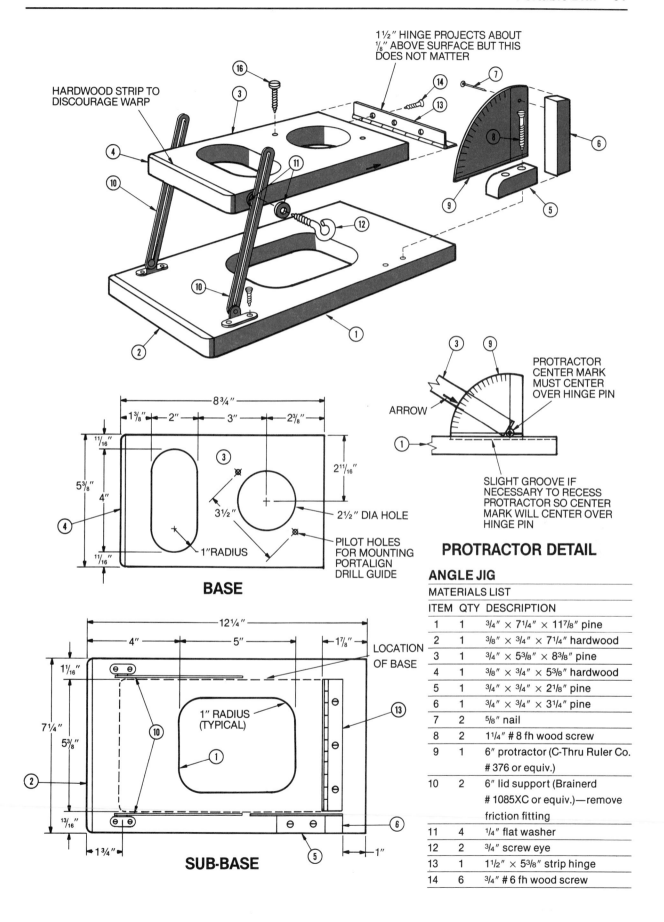

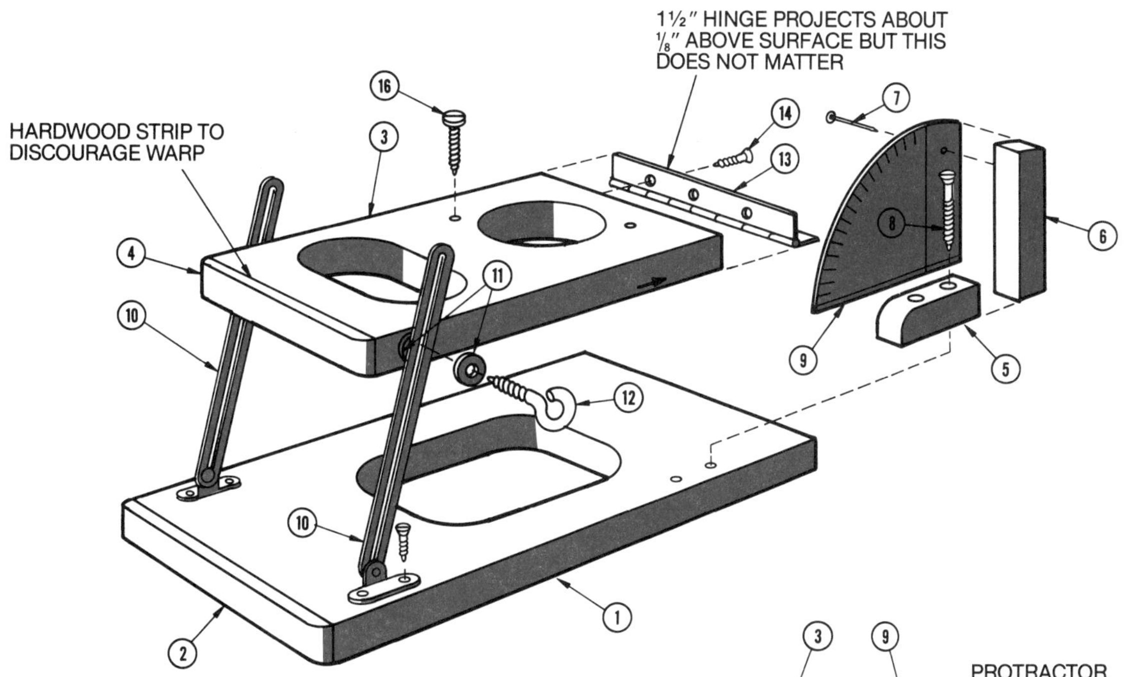
1 1/2" HINGE PROJECTS ABOUT 1/8" ABOVE SURFACE BUT THIS DOES NOT MATTER
HARDWOOD STRIP TO DISCOURAGE WARP
16
3
14
7
13
4
11
10
8
6
9
5
12
10
1
2

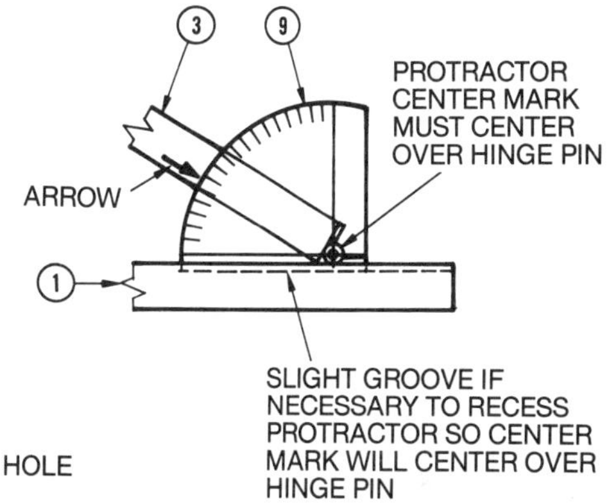
3
9
ARROW
PROTRACTOR CENTER MARK MUST CENTER OVER HINGE PIN
1
SLIGHT GROOVE IF NECESSARY TO RECESS PROTRACTOR SO CENTER MARK WILL CENTER OVER HINGE PIN

PROTRACTOR DETAIL

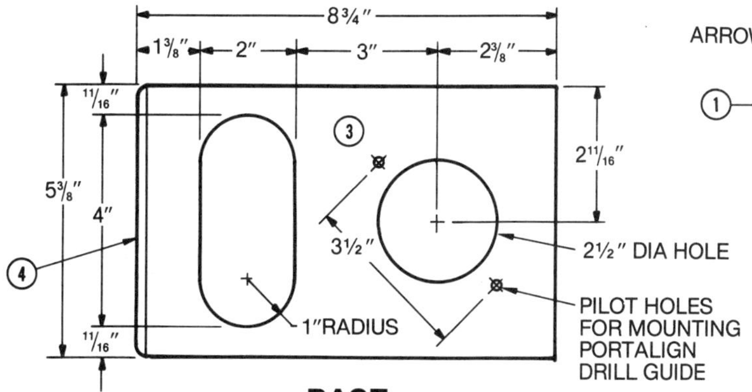
8 3/4"
1 3/8"
2"
3"
2 3/8"
11/16"
5 3/8"
4"
11/16"
3
2 11/16"
3 1/2"
2 1/2" DIA HOLE
4
1"RADIUS
PILOT HOLES FOR MOUNTING PORTALIGN DRILL GUIDE

BASE

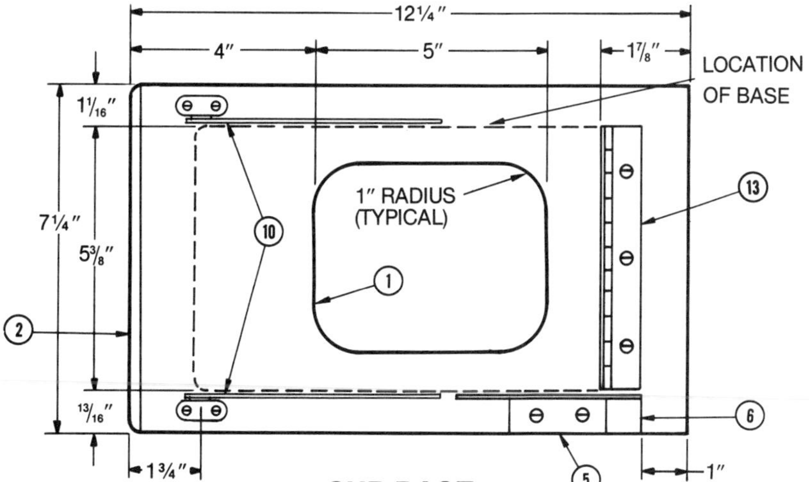
12 1/4"
4"
5"
1 7/8"
LOCATION OF BASE
1 1/16"
7 1/4"
5 3/8"
13/16"
1" RADIUS (TYPICAL)
10
1
13
2
6
5
1 3/4"
1"

SUB-BASE

ANGLE JIG

MATERIALS LIST

ITEM	QTY	DESCRIPTION
1	1	3/4" × 7 1/4" × 11 7/8" pine
2	1	3/8" × 3/4" × 7 1/4" hardwood
3	1	3/4" × 5 3/8" × 8 3/8" pine
4	1	3/8" × 3/4" × 5 3/8" hardwood
5	1	3/4" × 3/4" × 2 1/8" pine
6	1	3/4" × 3/4" × 3 1/4" pine
7	2	5/8" nail
8	2	1 1/4" # 8 fh wood screw
9	1	6" protractor (C-Thru Ruler Co. # 376 or equiv.)
10	2	6" lid support (Brainerd # 1085XC or equiv.)—remove friction fitting
11	4	1/4" flat washer
12	2	3/4" screw eye
13	1	1 1/2" × 5 3/8" strip hinge
14	6	3/4" # 6 fh wood screw

the protractor as this is the critical element in the jig. The center point at the base, where the 0-degree and 90-degree axes intersect, must coincide precisely with the center of the hinge pin. Cut the protractor down as shown here. Depending on the cross section of the hinge and the space between the zero base line and the bottom edge of the protractor, it may be necessary to cut a shallow groove in the sub-baseboard to permit recessing the protractor to achieve the required pivot point alignment.

Drill the screw holes for attaching the protractor bracket but *do not* attach the bracket. Align the protractor, then hold it in place by taping it to the edge of the tilt board. Now temporarily attach the bracket with screws and tape the protractor to the bracket. Remove the tape from the tilt board, then withdraw the screws holding the bracket. Remove the bracket together with the protractor taped in place. Join them with 5/8-inch nails. This procedure will enable you to mount the protractor to the bracket in perfect alignment; again, perfect alignment is paramount.

The tilt board is supported with two Brainerd lid supports No. 1085XC. This item is commonly available in hardware stores. It has a friction fitting which is of no use for this application. Discard the fitting and substitute a 3/4-inch screw eye with two washers to facilitate locking the tilt board.

Ink-mark an arrow index on the edge of the tilt board opposite the zero graduation on the protractor. (You can also use a press-type arrow which is available at art supply outlets.) Apply shellac to the entire jig (this will also protect the press-type arrow). Complete the jig by rubber-cementing a sheet of 220-grit abrasive paper to the bottom of the base to render it slip-proof.

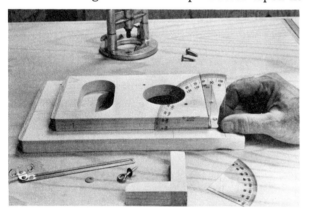

The protractor may have to be set into a groove, as was the case here, in order to obtain the necessary alignment of center point to hinge pin.

Close-up shows protractor center point in alignment with hinge pin center.

Portalign drill guide attaches to the tilt board with two screws. Elongated cutout in tilt board is a viewing port. The friction fitting of the lid support (pictured in the left foreground) is replaced with screw eye.

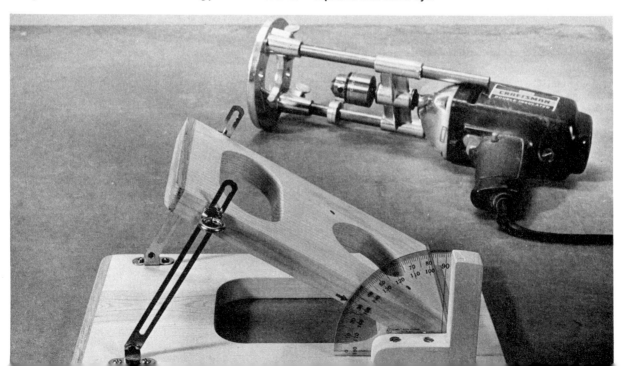

Self-Centering Punch

You can mark drilling centers quickly and accurately in the edges of boards with this handy self-centering punch. The centering pins and punch are made of 1/4-inch-diameter steel rod. Taper-point the punch by chucking it in a portable drill and work it against a spinning grinding wheel.

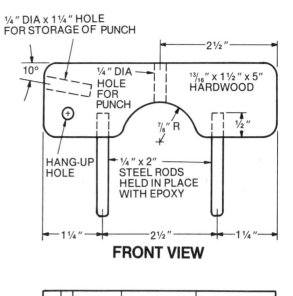

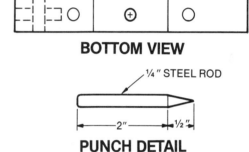

Centermark holes will be dead center on an edge when made with this punch. Hole in the end stores the punch; hole in the side is for hanging it up when not in use.

Angle Drilling Guide

You can drill an angled hole with reasonable accuracy in the following manner: bore a straight hole edgewise through a scrap of 2 × 4. Cut through the block at the desired angle, then make a semicircular cutout centered over the hole. Insert the drill bit into the block and view through the opening to center the bit on the mark. Hold the block firmly in place while drilling.

This is how the inclined block is used to drill angled holes.

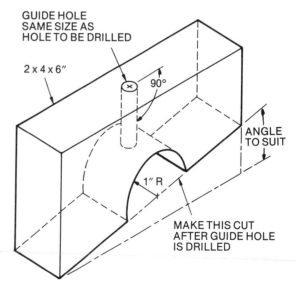

Drill Stand

The drill stand is a simple, low-cost, commercially available accessory that is useful for a variety of jobs in a shop with limited equipment. It will enable you to utilize your drill as a stationary arbor. The applications illustrated show but a few of the possibilities.

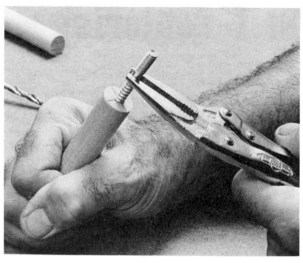

Spindle shaping on a small scale is possible. Wood thread portion of a hanger bolt is inserted in dowel stock.

Buffing wheel can be chucked into the drill to simplify polishing chores.

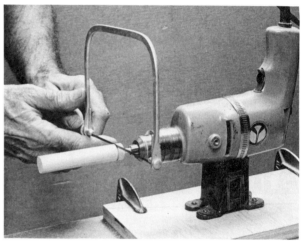

Saw is used to make a shoulder cut.

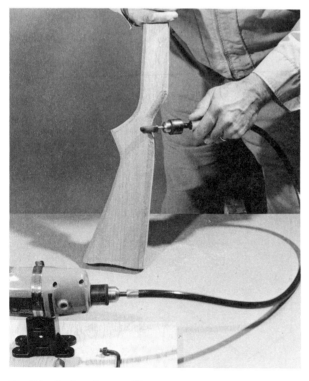

Flexible shaft and rotary file are teamed up to make contouring work easier.

Rasp and file are used to reduce the diameter. This project is a replacement handle for a tool.

Horizontal Boring Jig

A horizontal boring capability enables you to make holes accurately in oversize workpieces, especially useful for drilling holes in the edges of large boards.

This jig utilizes a Portalign drill guide mounted on a simple but effective screw-controlled elevating slide that permits precise positioning of the drill. Large holes in the platform provide clamp access for narrow boards. Oversize panels or crosswise pieces that do not coincide with the clamping holes can be clamped to the ledges alongside the backboard. Ledges along the bottom allow clamping of the unit to a workbench when necessary.

The parts can be made of 1×12 pine or $3/4$-inch plywood, but the elevating nut bracket should be made of hardwood. This nut is encased in a hexshaped recess. An easy way to make the recess for the nut is to construct the bracket in two parts as shown here. Cut the hexagonal outline with a scroll or coping saw, then glue the bottom block to the sawed-out piece. A bead of quick-setting epoxy around the edge of the nut will lock it permanently.

The channel for the sliding drill board must be made with care, the fit neither too tight nor loose. Seal the wood with shellac and apply paste wax to the contact surfaces of the drill board for easy sliding.

MATERIALS LIST		
ITEM	QTY	DESCRIPTION
1	1	$3/4'' \times 11'' \times 29^3/4''$ pine
2	2	$3/4'' \times 2^3/4'' \times 29^3/4''$ pine
3	2	$3/4'' \times 1^3/4'' \times 30''$ pine
4	2	$3/4'' \times 1^3/4'' \times 4''$ pine
5	1	$3/4'' \times 3^1/2'' \times 11''$ pine
6	1	$3/4'' \times 1^3/4'' \times 11''$ pine
7	1	$3/4'' \times 11'' \times 11''$ pine
8	2	$1^1/8'' \times 1^1/4'' \times 11''$ pine
9	1	$3/4'' \times 3'' \times 11''$ pine
10	1	$3/4'' \times 5^3/4'' \times 7^1/4''$ pine
11	1	$1'' \times 4''$ dowel
12	1	$3/8'' \times 2''$ dowel
13	1	$1/8'' \times 1^1/4''$ rh machine screw
14	1	$1/8''$ hex nut
15	2	$3/8''$ hex nut
16	2	cotter pin
17	2	$3/8''$ flat washer
18	1	$3/8'' \times 9''$ threaded steel rod
19	1	$1/2'' \times 1^1/4'' \times 2^1/2''$ hardwood
20	1	$13/16'' \times 1^1/4'' \times 2^1/2''$ hardwood
21	2	$1''$ # 8 fh wood screw
22	2	$3/8'' \times 5^3/4''$ drawer slide tape
23	2	$1''$ # 14 pan-hd sheet-metal screw

Using the jig to bore into the end of a board. The crank handle is turned to raise or lower the drill.

Backboard clamp ledges are utilized when the work cannot be clamped on the platform.

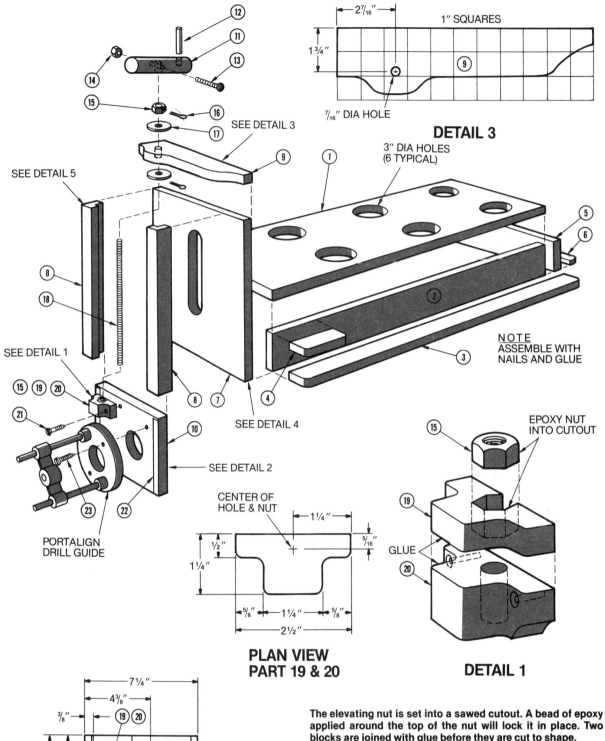

2⁷⁄₁₆″

1″ SQUARES

1³⁄₄″

⑨

⁷⁄₁₆″ DIA HOLE

DETAIL 3

⑫
⑪
⑬
⑭
⑮
⑯
⑰

SEE DETAIL 3

SEE DETAIL 5

⑨

①

3″ DIA HOLES
(6 TYPICAL)

⑤
⑥

⑧
⑱

②

③

<u>NOTE</u>
ASSEMBLE WITH
NAILS AND GLUE

SEE DETAIL 1

⑮ ⑲ ⑳
㉑

⑧ ⑦
④

SEE DETAIL 4

㉓ ㉒

⑩

SEE DETAIL 2

PORTALIGN
DRILL GUIDE

CENTER OF
HOLE & NUT

1¹⁄₄″

¹⁄₂″   ⁵⁄₁₆″

1¹⁄₄″

⁵⁄₈″   1¹⁄₄″   ⁵⁄₈″

2¹⁄₂″

**PLAN VIEW
PART 19 & 20**

⑮

EPOXY NUT
INTO CUTOUT

⑲

GLUE

⑳

DETAIL 1

7¹⁄₄″

4³⁄₈″

³⁄₈″

⑲ ⑳

2⁷⁄₈″

5³⁄₄″

⑩

1³⁄₄″ DIA
HOLE

2⁷⁄₈″

㉒

DETAIL 2

The elevating nut is set into a sawed cutout. A bead of epoxy
applied around the top of the nut will lock it in place. Two
blocks are joined with glue before they are cut to shape.

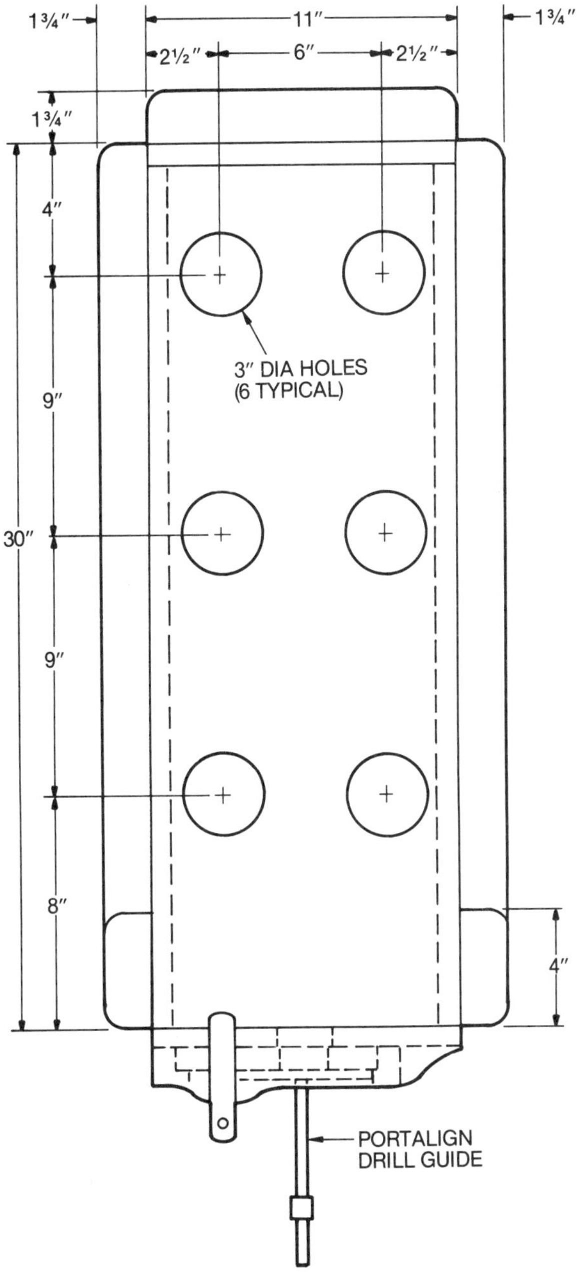

1¾" 11" 1¾"
2½" 6" 2½"

1¾"

4"

3" DIA HOLES
(6 TYPICAL)

9"

30" 9"

8"

4"

PORTALIGN
DRILL GUIDE

PLAN VIEW

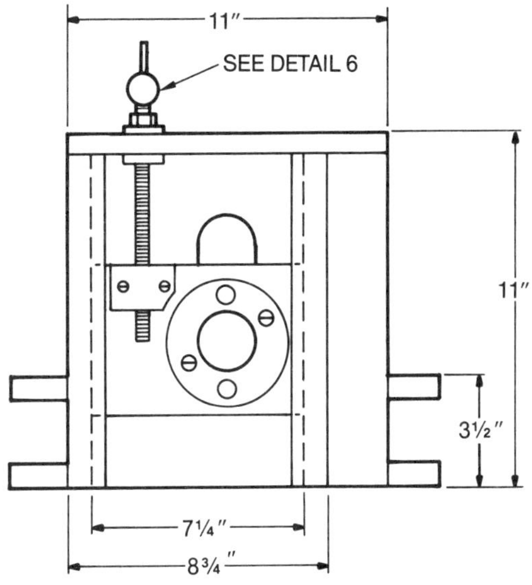

FRONT VIEW

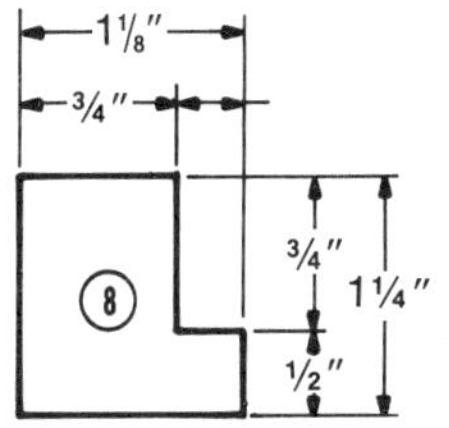

11"

SEE DETAIL 6

11"

3½"

7¼"

8¾"

DETAIL 5

1⅛"
¾"

⑧

¾" 1¼"
½"

DETAIL 6

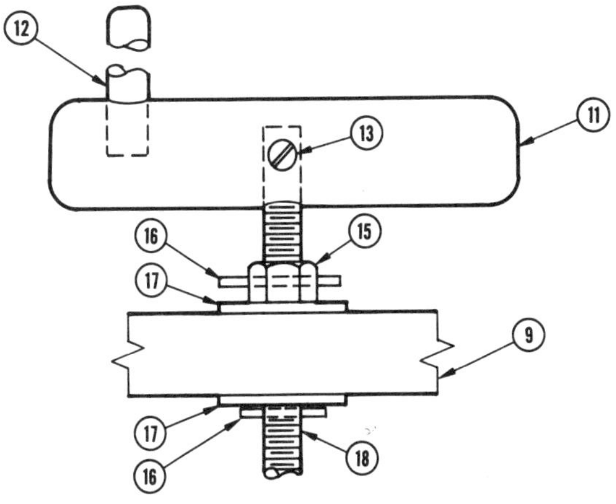

⑫

⑬

⑪

⑯ ⑮

⑰

⑨

⑰

⑯ ⑱

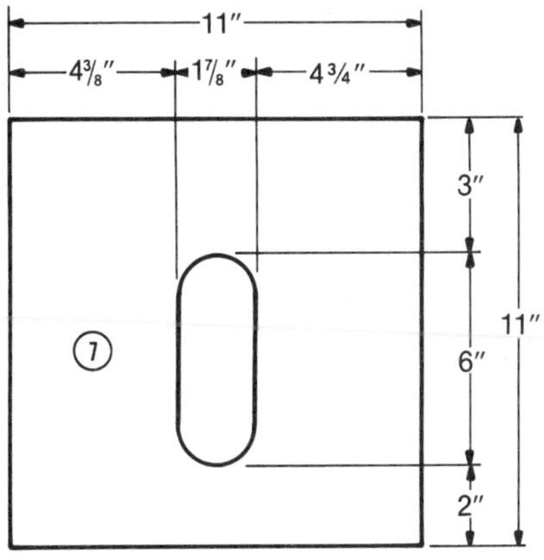

DETAIL 4

11"
4⅜" 1⅞" 4¾"

3"

11"

⑦ 6"

2"

8. DRILL PRESS

Auxiliary Tilt Table

This jig is invaluable if your drill press lacks a tilting table. And it is even advantageous for a drill press that does have a tilting table because of the ease with which it permits precise angular adjustments. Built-in tilt tables are adjusted by sighting the drill bit against a protractor or a T bevel. This is something of a trial-and-error procedure—one that frequently results in inaccuracies. The direct-reading angle scale on this jig avoids any guesswork.

Setup is quick and simple. The base of the unit slips into place with a snug fit, thus eliminating the need for bolting or clamping.

An adjustable, sliding support arm holds table at the desired angle from zero at 45 degrees. The settings are read against a protractor that swings with the table. Four diagonal slots in the top provide clamp access for holding small work. Large work is clamped around the perimeter. The rectangular slot in the center accepts scrap blocks of wood; these serve as disposable back-ups for through drilling.

The base can be made of pine or fir plywood, but the tilt top should be made of hardwood plywood.

This jig permits angular drilling with great precision.

MATERIALS LIST		
ITEM	QTY	DESCRIPTION
1	1	3/4" × 12" × 20" ash veneer plywood
2	1	3/4" × 11 1/2" × 12 1/4" pine
3	1	3/4" × 4 3/4" × 11 1/2" pine
4	2	3/4" × 4" × 10 3/4" pine
5	1	3/4" × 3 1/4" × 10" pine
6	2	3/4" × 2 1/8" × 4" pine
7	1	3/4" × 3" × 18" pine
8	2	3/8" × 3/4" × 4" pine
9	1	1 1/2" × 12" strip hinge (with twelve 3/4" # 6 fh wood screws)
10	1	6" protractor (C-Thru Ruler Co. # 376 or equiv.)
11	1	1/8" × 5/8" × 3" pine
12	16	2" finishing nail
13	1	3/8" × 2" carriage bolt
14	1	1 1/2" fender washer
15	1	3/8" wing nut
16	6	1" nail
17	8	1 1/4" finishing nail

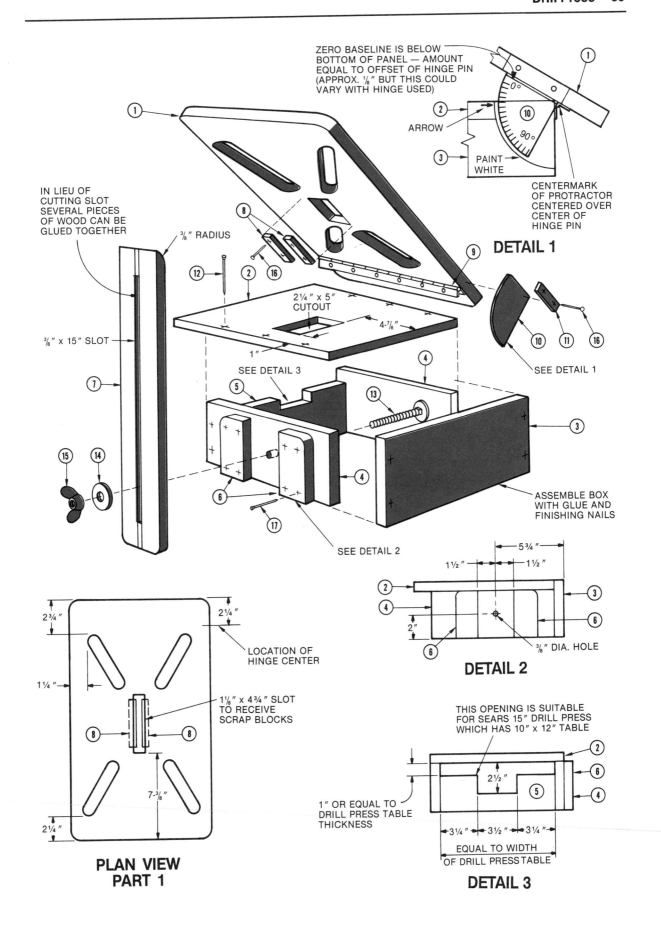

ZERO BASELINE IS BELOW BOTTOM OF PANEL — AMOUNT EQUAL TO OFFSET OF HINGE PIN (APPROX. ⅛″ BUT THIS COULD VARY WITH HINGE USED)

ARROW

PAINT WHITE

CENTERMARK OF PROTRACTOR CENTERED OVER CENTER OF HINGE PIN

DETAIL 1

IN LIEU OF CUTTING SLOT SEVERAL PIECES OF WOOD CAN BE GLUED TOGETHER

⅜″ RADIUS

⅜″ x 15″ SLOT

2¼″ x 5″ CUTOUT

4-⅞″

1″

SEE DETAIL 3

SEE DETAIL 1

SEE DETAIL 2

ASSEMBLE BOX WITH GLUE AND FINISHING NAILS

2¾″

2¼″

1¼″

LOCATION OF HINGE CENTER

1⅛″ x 4¾″ SLOT TO RECEIVE SCRAP BLOCKS

7-⅜″

2¼″

PLAN VIEW PART 1

5¾″

1½″   1½″

2″

⅜″ DIA. HOLE

DETAIL 2

THIS OPENING IS SUITABLE FOR SEARS 15″ DRILL PRESS WHICH HAS 10″ x 12″ TABLE

2½″

1″ OR EQUAL TO DRILL PRESS TABLE THICKNESS

3¼″   3½″   3¼″

EQUAL TO WIDTH OF DRILL PRESS TABLE

DETAIL 3

The unit is sized for a Sears 15-inch drill press, which has a 10-by-12-inch table. Alter the dimensions to suit other drill presses.

Construct the base box so the inside width exactly matches the width of the press table in order to obtain a good sliding fit without play. The rear cross member should be fitted so that it bears against the bottom of the table when the box is in position.

As described for the portable drill Angle Jig (page 60), the protractor must be attached so its center point aligns directly opposite the hinge pivot pin center. Unless a common axis is obtained the angular settings of the table will be in error. The hinge pin center will be offset a small amount below the bottom surface of the tilt table. When attaching the protractor, be certain to keep its base parallel to the surface of the table. This will locate the zero graduation below the bottom surface of the table. Accordingly, the index arrow should be located a like amount below the interface.

Paint a white patch behind the protractor for good visibility, and use a press-type arrow or ink to mark the index. Seal the wood with several thin coats of shellac or use a clear satin top coat finish.

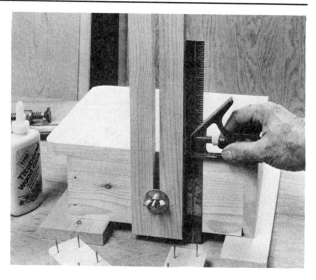

Support arm is locked in a trued position before the guide blocks are attached.

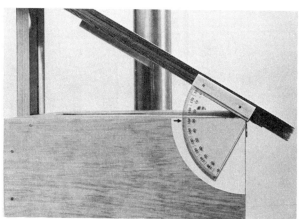

Close-up of the adjustment scale. The inside row of numbers applies with this particular protractor. The arrow point indicates the set angle of the table.

Replaceable scrap wood insert serves as back-up for through drilling to prevent splintering of work when the bit exits. The insert rests on two small cleats under the rectangular slot.

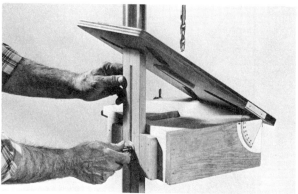

Wing nut locks the support arm in the desired position.

Table Alignment Guide

Centering the drill-press table each time it is lowered or raised must be done by visual alignment of the chuck center over the clearance hole in the center of the table because there are no other positive points of reference for centering. This explains why many drill-press tables have pockmarks and scars in an arc adjacent to the center hole.

A misalignment when using a metal-cutting drill bit merely results in another scar in the table. But the expensive wood-boring bit that misses the center hole and strikes the steel table will be damaged.

The problem can be solved by marking a clearly visible centerline down the length of the column and a small reference mark at the rear center of the table. This is accomplished by scribing a line with a carbide-tipped scriber of the type used for cutting laminates and plastic sheet materials. A diamond-pointed scriber may be used. The scribed line should then be filled with white paint.

Clamp a length of angle iron to the column and run the tool along the edge, making several passes until a shallow groove is obtained. This will cause some burring; remove it with a smooth file. Apply white paint to the groove, then wipe over the surface of the column with a tight pad of cloth dampened with the appropriate solvent to remove excess paint. A visible white line will remain in the depression. Make the index mark in the table in the same manner.

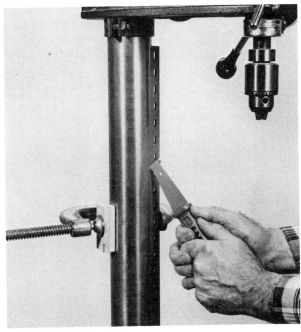

Clamped angle iron guides the carbide-tipped scriber. A few firm passes are required.

White line on the column and table center will enable you to center the table quickly and easily.

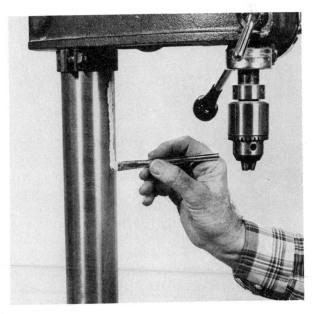

Paint over the scribed line, then wipe off the excess on the surface before it dries.

Column Fence

A column fence is a useful fixture for the drill press because it will enable you to do precision hole drilling in the ends of relatively long boards. The length of the boards that can be accommodated is limited to the maximum space between the drill bit and the lowered table on which the bottom of the work must rest. Since column fences are not commercially available, you have no choice but to build your own. The one shown here will serve you well.

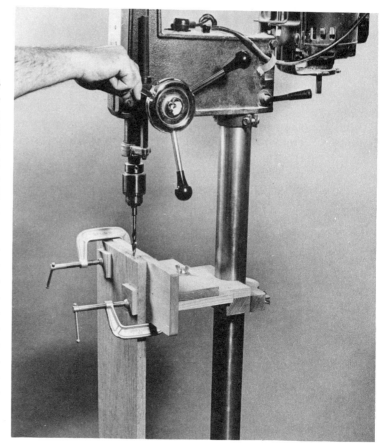

Column fence in use. Bottom of the work must rest on the lowered table for solid footing.

MATERIALS LIST

ITEM	QTY	DESCRIPTION
1	1	$3/4'' \times 5'' \times 12''$ pine
2	1	$3/4'' \times 4^{1/2}'' \times 8^{1/2}''$ pine
3	2	$3/4'' \times 3/4'' \times 6''$ pine
4	1	$3/4'' \times 6'' \times 7''$ pine
5	1	$2 \times 4 \times 7''$
6	1	$2 \times 4 \times 7''$
7	1	$3/8'' \times 2''$ carriage bolt
8	2	$3/8'' \times 4''$ carriage bolt
9	3	$3/8''$ flat washer
10	3	$3/8''$ wing nut
11	6	2'' finishing nail
12	4	$1^{1/4}''$ finishing nail

Uniform Spacing Jig

This jig will prove most helpful for drilling a series of holes with uniform spacing in a straight line. It is set up in the following manner: mark the centers for the first two holes in the series. Bore a hole in a small block and put it aside momentarily. Position a wood fence and clamp it to the table, then drill the first hole in the work. Shift the work so the center mark for the second hole is in line with the drill bit. Insert a pin through the small block and into the hole in the work. Nail the pin block to the fence.

The work is shifted until the stop pin drops into the previously drilled hole.

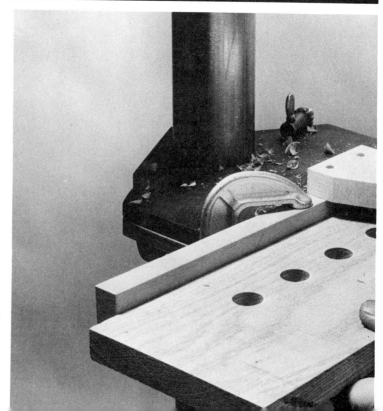

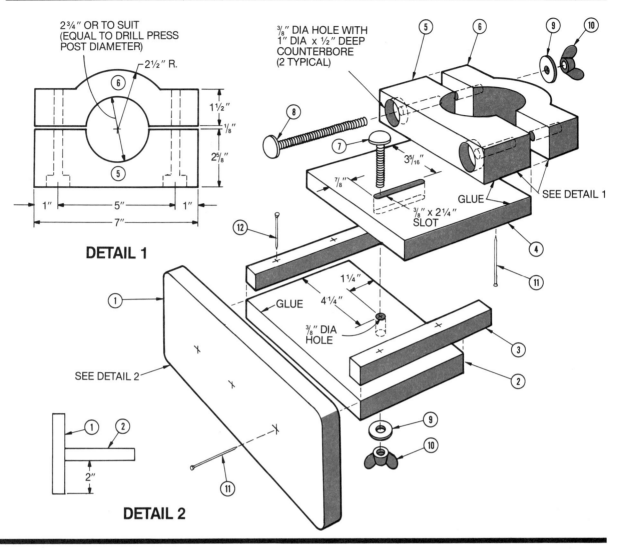

2¾" OR TO SUIT
(EQUAL TO DRILL PRESS
POST DIAMETER)

2½" R.

⑥

⑤

1½"

⅛"

2⅝"

1"

5"

1"

7"

DETAIL 1

⅜" DIA HOLE WITH
1" DIA x ½" DEEP
COUNTERBORE
(2 TYPICAL)

⑤

⑥

⑨

⑩

⑧

⑦

3⁵⁄₁₆"

⅞"

GLUE

SEE DETAIL 1

⅜" x 2¼"
SLOT

④

⑪

⑫

①

GLUE

1¼"

4¹⁄₄"

⅜" DIA
HOLE

③

②

SEE DETAIL 2

①

②

2"

⑨

⑩

⑪

DETAIL 2

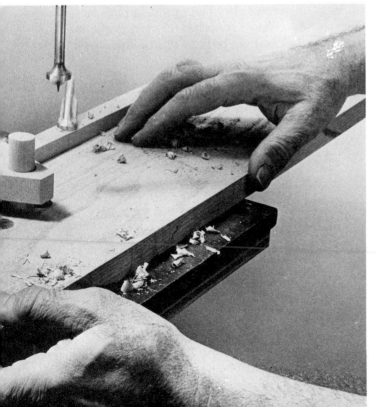

Now continue to drill the holes by simply shifting the work until the pin drops into the previously drilled hole. Notice that the pin can be a dowel, a steel rod, a screw, a nail—any item that is exactly the size of the hole.

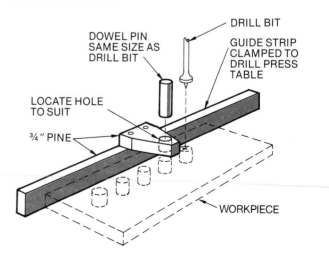

DOWEL PIN
SAME SIZE AS
DRILL BIT

DRILL BIT

GUIDE STRIP
CLAMPED TO
DRILL PRESS
TABLE

LOCATE HOLE
TO SUIT

¾" PINE

WORKPIECE

Pocket Hole Jig

Pocket holes provide a simple method for attaching rails to table tops and the like by permitting screws to be driven at a steep angle through the edge of the rail.

The jig holds the work at a 15-degree angle, and the pilot block allows the bit to enter the wood without skipping about. The jig is clamped to the drill-press table, and the work is shifted to position for the succeeding hole. A spur bit works best for this operation.

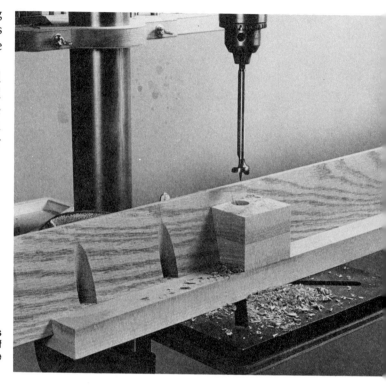

Using the jig to cut pocket holes. When all the pocket holes have been made, the jig is shifted to move the pilot block out of the way. The holes for the screw shank are then drilled at the same angle.

Accessory Chest

The numerous accessories that are used with the drill press can be stored conveniently and within easy reach in this chest which is kept in the generally idle space at the base of the press. Four diagonal cleats on the bottom drop into the slots in the drill base to prevent the chest from slipping off. When the space below is needed for working with long pieces the chest can be lifted off readily.

This handy chest features two trays that slide sideways to allow access to the bottom compartment large enough for storing bulkier items.

MATERIALS LIST		
ITEM	QTY	DESCRIPTION
1	1	1/2" × 11 3/4" × 16 1/2" poplar
2	1	1/2" × 1 1/8" × 16 1/2" poplar
3	2	1/2" × 9 1/2" × 16 1/2" poplar
4	2	1/2" × 11 7/8" × 9" poplar
5	1	1/2" × 11 7/8" × 16 1/2" poplar
6	4	1/2" × 1/2" × 15 1/2" pine
7	2	1 1/2" × 1 1/2" butt hinge
8	2	1/4" × 1 1/8" × 11 13/16" pine
9	2	1/4" × 1 1/8" × 7 1/2" pine
10	2	1/4" × 8" × 11 13/16" Luan plywood
11	2	1/4" × 2 1/2" × 11 13/16" pine
12	2	1/4" × 2 1/2" × 7 1/2" pine
13	4	1/2" pine (size to suit)

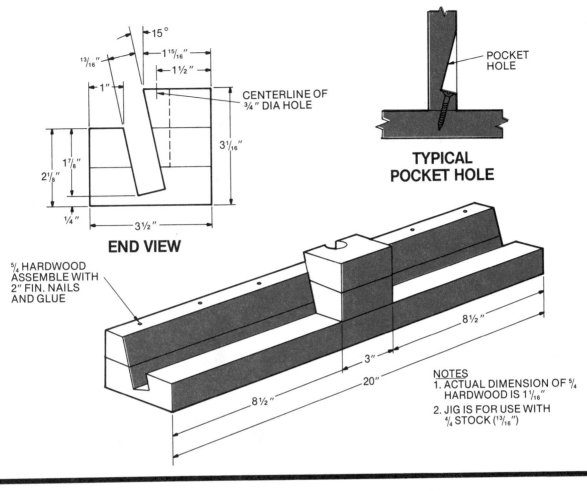

15°

$1\frac{15}{16}''$

$\frac{13}{16}''$

$1\frac{1}{2}''$

$1''$

CENTERLINE OF
¾" DIA HOLE

$3\frac{1}{16}''$

$1\frac{7}{8}''$

$2\frac{1}{8}''$

¼"

$3\frac{1}{2}''$

END VIEW

POCKET
HOLE

**TYPICAL
POCKET HOLE**

⁵⁄₄ HARDWOOD
ASSEMBLE WITH
2" FIN. NAILS
AND GLUE

$8\frac{1}{2}''$

3"

20"

$8\frac{1}{2}''$

NOTES
1. ACTUAL DIMENSION OF ⁵⁄₄
   HARDWOOD IS $1\frac{1}{16}''$
2. JIG IS FOR USE WITH
   ⁴⁄₄ STOCK ($\frac{13}{16}''$)

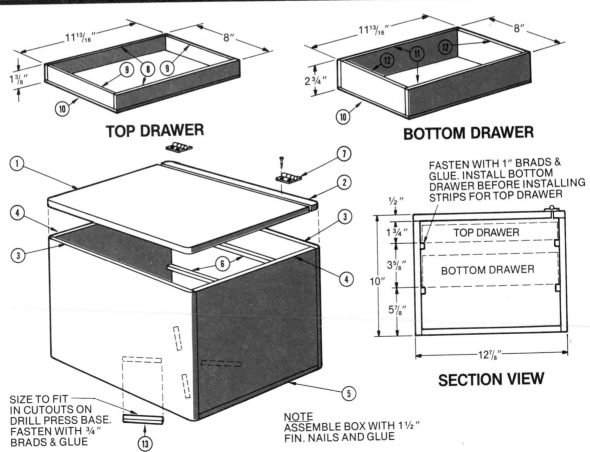

$11\frac{13}{16}''$

8"

$1\frac{3}{8}''$

⑨ ⑧ ⑨

⑩

TOP DRAWER

$11\frac{13}{16}''$

8"

⑫ ⑪ ⑫

$2\frac{3}{4}''$

⑩

BOTTOM DRAWER

FASTEN WITH 1" BRADS &
GLUE. INSTALL BOTTOM
DRAWER BEFORE INSTALLING
STRIPS FOR TOP DRAWER

½"

$1\frac{3}{4}''$

TOP DRAWER

$3\frac{5}{8}''$

BOTTOM DRAWER

10"

$5\frac{7}{8}''$

$12\frac{7}{8}''$

SECTION VIEW

① ② ③ ④ ⑤ ⑥ ⑦

SIZE TO FIT
IN CUTOUTS ON
DRILL PRESS BASE.
FASTEN WITH ¾"
BRADS & GLUE

⑬

NOTE
ASSEMBLE BOX WITH $1\frac{1}{2}''$
FIN. NAILS AND GLUE

Auxiliary Table

Drill-press tables generally prove to be a bit too small for typical woodworking operations, but this problem can be solved with a slip-on oversize table. The one shown here features an adjustable fence that is useful for multiple-hole drilling and routing operations. A square cutout in the center accepts disposable back-up blocks that permit splinter-free through drilling. Inserts with round cutouts are used to house the bottoms of drum-sanding attachments. An accessory spacer block is shown in the drawing. This is used to provide extra bottom clearance when it is desirable to utilize the full length of a sanding drum.

MATERIALS LIST		
ITEM	QTY	DESCRIPTION
1	1	3/4" × 21" × 24" plywood
2	1	3/4" × 2½" × 20" plywood
3	1	3/4" × 2½" × 20" plywood
4	2	3/4" × 4½" × 4½" pine
5	2	½" × 1¼" × 14¼" poplar
6	1	½" × 1¼" × 12½" poplar
7	2	5/16" × 5" carriage bolt
8	4	1½" fender washer
9	4	5/16" wing nut
10	2	5/16" × 2" carriage bolt

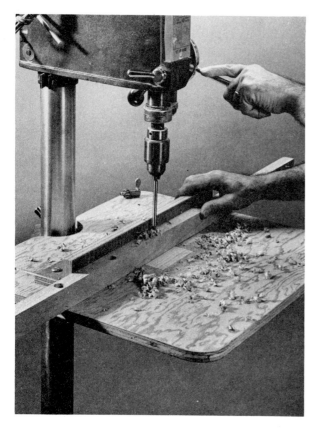

Oversize table is an asset for most woodworking chores on the drill press. Adjustable fence is useful for multiple drilling.

Sanding-drum bottom sets into a hole in the insert block.

This insert is used for drum sanding. Several sizes should be kept on hand for use with drums of different diameters. The block in the foreground is used for drill back-up.

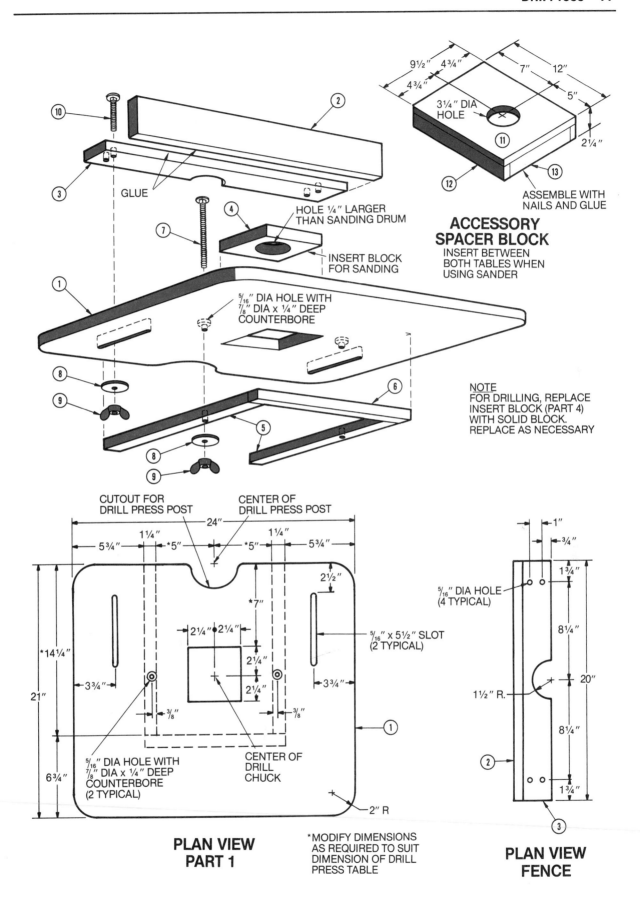

9½" 4¾" 7" 12"

4¾" 5"

3¼" DIA
HOLE

⑪

2¼"

⑫ ⑬

ASSEMBLE WITH
NAILS AND GLUE

**ACCESSORY
SPACER BLOCK**
INSERT BETWEEN
BOTH TABLES WHEN
USING SANDER

②

⑩

③

GLUE

HOLE ¼" LARGER
THAN SANDING DRUM

④

⑦

INSERT BLOCK
FOR SANDING

①

⁵⁄₁₆" DIA HOLE WITH
⁷⁄₈" DIA x ¼" DEEP
COUNTERBORE

⑧

⑨

⑥

⑤

⑧

⑨

NOTE
FOR DRILLING, REPLACE
INSERT BLOCK (PART 4)
WITH SOLID BLOCK.
REPLACE AS NECESSARY

CUTOUT FOR
DRILL PRESS POST

CENTER OF
DRILL PRESS POST

24"

5¾" 1¼" *5" *5" 1¼" 5¾"

2½"

*7"

⁵⁄₁₆" x 5½" SLOT
(2 TYPICAL)

*14¼"

2¼" 2¼"

2¼"

21"

3¾" 3¾"

2¼"

3/8" 3/8"

①

⁵⁄₁₆" DIA HOLE WITH
⁷⁄₈" DIA x ¼" DEEP
COUNTERBORE
(2 TYPICAL)

CENTER OF
DRILL
CHUCK

6¾"

2" R

**PLAN VIEW
PART 1**

*MODIFY DIMENSIONS
AS REQUIRED TO SUIT
DIMENSION OF DRILL
PRESS TABLE

1"

¾"

1¾"

⁵⁄₁₆" DIA HOLE
(4 TYPICAL)

8¼"

20"

1½" R.

8¼"

②

1¾"

③

**PLAN VIEW
FENCE**

Turning Jig

In a broad sense the drill press can be likened to a lathe standing on end. Of course it lacks a tailstock and tool rest, but if you would like to try your hand at making small turnings without the expense of buying a lathe you can do so on the drill press—use the improvised tailstock and tool rest shown.

The tailstock is made with three pieces of hardwood, a ball bearing ($^1/_4$ inch thick by $^3/_4$ inch diameter × $^1/_4$ inch bore), and a $^1/_4$–20 by $1^1/_2$-inch bolt with nut.

Use a brad point bit to bore a $^3/_4$-inch hole, $^1/_4$ inch deep, into the "tailstock" block. Counterbore an additional $^1/_4$ inch deep with a $^9/_{16}$-inch flat bit. This results in a clearance hole for the bolt head and a shoulder to support the bearing. Put the bolt through the bearing and secure with the nut, then tap the assembly into the hole. The bearing should fit snug so that only the bolt will spin.

A $^3/_8$-by-12-inch pipe threaded into a pipe flange mounted on a block of wood serves as the tool rest.

Bore a $^1/_4$-inch centered hole 1 inch deep into the bottom of a turning blank. Bore a centered pilot hole in the top of the blank to receive the wood thread portion of a $^1/_4$–20 by 2-inch hanger bolt. Thread a nut on the other end of the hanger bolt and use a wrench to insert the bolt into the wood.

A bearing and bolt make up the tailstock center. Bore the larger hole with the brad point bit first. This will preserve the center for the flat bit.

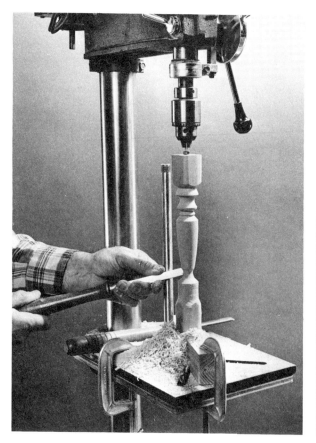

Small turnings of excellent quality can be made using this turning jig on the drill press.

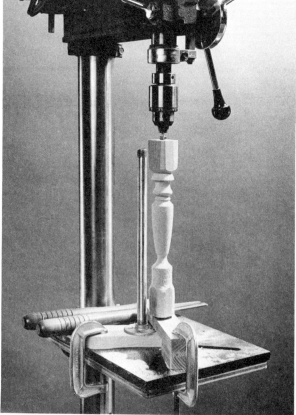

The wood chips have been cleared away here to show how the tool-rest flange fits under the tailstock support bracket.

Secure the hanger bolt projection in the drill chuck, insert the bottom of the blank over the tail-stock bolt, then clamp the block to the drill-press table. Position the tool rest close to the work and clamp it in place.

Using a basic set of turning chisels (which can be purchased fairly inexpensively), aim the appropriate chisel so the point is about 1/16 inch to the right of center.

This arrangement is suitable for work about 1 1/2 inches in diameter and about 12 inches long. For larger work a 1/2-inch pipe should be used for the tool rest. For a source of supply for the bearing look in the Yellow Pages under Bearings.

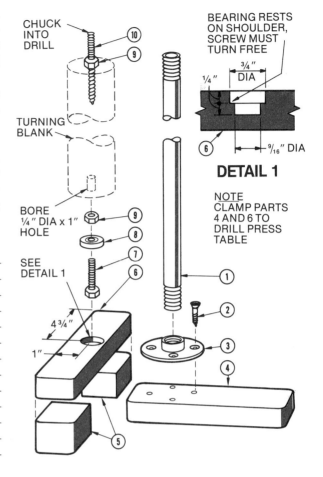

MATERIALS LIST

ITEM	QTY	DESCRIPTION
1	1	3/8" × 12" pipe (or length as required)
2	4	3/4" # 14 fh wood screw
3	1	3/8" floor flange
4	1	3/4" × 2 3/4" × 6" hardwood
5	2	1 3/16" × 1 1/2" × 2 1/2" hardwood
6	1	1 3/16" × 1 1/2" × 9 1/2" hardwood
7	1	1/4"–20 × 1 1/2" bolt
8	1	1/4"-thick × 3/4"-dia. × 1/4"-bore ball bearing
9	2	1/4" hex nut
10	1	1/4" × –20 × 2" hanger bolt

V Block for Rounds

Make two 45-degree bevel cuts to form a V notch in a piece of wood and attach this to a baseboard to permit clamping to the drill-press table.

This jig will permit drilling centered holes through dowels or round metal stock. To set it up, align the center of the notch with the drill bit. When a series of holes is required, drive a nail part of the way through the previously drilled hole into the V block; this will hold the work so you will obtain uniformly centered holes.

The V block aids drilling centered holes through dowels and round metal stock.

Drill Bit Rack

Keep your most active drill bits within easy reach in this rack that mounts on the drill-press column.

To make the wraparound base and clamp, bore a hole of the same diameter as the column in a plywood panel. Then make a lengthwise saw cut through the hole center to yield the two sections. The stock removed by the saw kerf reduces the diameter of the two semicircles sufficiently to permit tight clamping.

Drill bits are always within easy reach when stored in this rack.

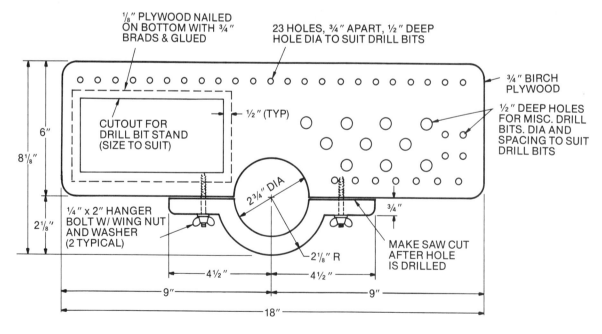

PLAN VIEW

Catchall Tray

A number of odds and ends usually accumulate while work progresses on the drill press. You can keep them handy but safely off the drill-press table with a catchall tray.

This one is shaped to fit the rear of the Sears 15-inch press. It attaches with a bolt threading into the hole that serves to hold an accessory tilt table on this particular machine. The tray can be shaped and adapted to fit almost any drill press.

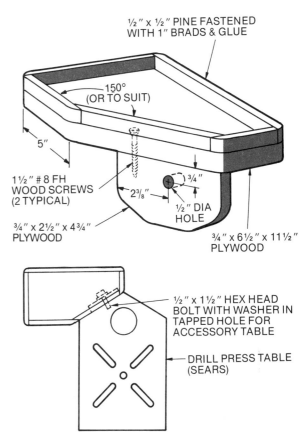

½" x ½" PINE FASTENED WITH 1" BRADS & GLUE

150° (OR TO SUIT)

5"

1½" # 8 FH WOOD SCREWS (2 TYPICAL)

¾" x 2½" x 4¾" PLYWOOD

2³⁄₈"

¾"

½" DIA HOLE

¾" x 6½" x 11½" PLYWOOD

½" x 1½" HEX HEAD BOLT WITH WASHER IN TAPPED HOLE FOR ACCESSORY TABLE

DRILL PRESS TABLE (SEARS)

PLAN VIEW

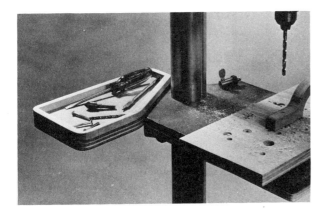

Keep things handy but out of the way with this simple tray.

Table Bottom Pad

Because most drill-press table castings have ribbed bottoms, it is often difficult to place a C clamp exactly where it is needed. Solve the problem by attaching a flush-fitting panel to provide an even surface for the clamp jaw.

Trace the outline and cutouts to suit the configuration of your own machine and cut the panel from a piece of ½-inch hardwood plywood. Drill and tap two holes in the table and attach the panel with flat head screws driven up through the bottom.

This addition will make it easier to apply clamps to the drill-press table.

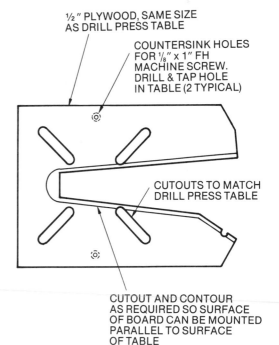

½" PLYWOOD, SAME SIZE AS DRILL PRESS TABLE

COUNTERSINK HOLES FOR ⅛" x 1" FH MACHINE SCREW. DRILL & TAP HOLE IN TABLE (2 TYPICAL)

CUTOUTS TO MATCH DRILL PRESS TABLE

CUTOUT AND CONTOUR AS REQUIRED SO SURFACE OF BOARD CAN BE MOUNTED PARALLEL TO SURFACE OF TABLE

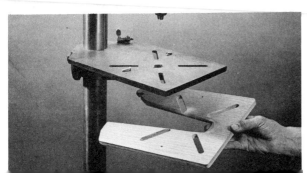

Retractable Dolly

While the drill press is basically a stationary tool, there are times—particularly in crowded quarters—when it is necessary to move it about to gain clearance for oversize workpieces.

Mounted with bolts on this dolly, the heavy machine can be moved safely and without the danger of causing a strained back. The levers are shifted to raise and lower the dolly so that either the wheels or blocks are in contact with the floor.

Join two pieces of ¾-inch plywood to make the platform and size it large enough so the casters will be located at least several inches outside the perimeter of the machine's base.

ITEM	QTY	DESCRIPTION
		MATERIALS LIST
1	2	¾″ × 18″ × 24″ plywood
2	2	1½″ × 2¼″ × 18″ (cut from 2 × 4 stock)
3	2	¾″ × 3″ × 10″ plywood
4	4	2½″ # 14 fh wood screws
5	4	plate swivel casters with 1½″-dia. wheel
6	16	1¼″ # 10 rh wood screws
7	3	⅜″ × 3″ hex-hd bolts
8	3	⅜″ flat washers
9	3	⅜″ hex nut
10	2	1½″ × 17″ strip hinge
11	36	1″ # 6 fh wood screws
12	4	3-prong steel glides

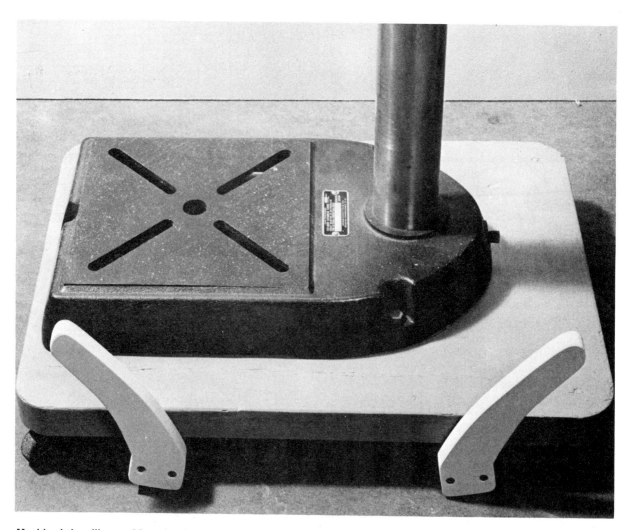

Machine is in rolling position when levers are spread apart.

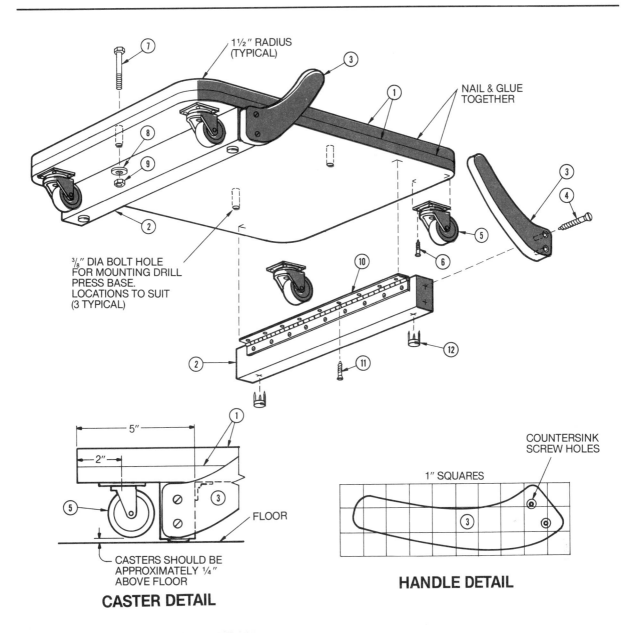

1½″ RADIUS
(TYPICAL)

NAIL & GLUE
TOGETHER

³⁄₈″ DIA BOLT HOLE
FOR MOUNTING DRILL
PRESS BASE.
LOCATIONS TO SUIT
(3 TYPICAL)

— 5″ —

— 2″ —

FLOOR

CASTERS SHOULD BE
APPROXIMATELY ¼″
ABOVE FLOOR

CASTER DETAIL

COUNTERSINK
SCREW HOLES

1″ SQUARES

HANDLE DETAIL

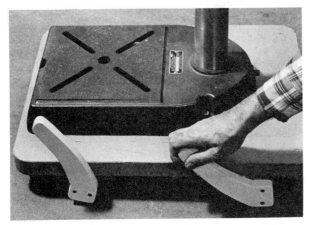

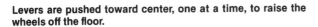

Levers are pushed toward center, one at a time, to raise the
wheels off the floor.

Dolly is now in the stationary position; wheels are up, support
blocks are down.

9. ROUTER

Slab Surfacing Jig

Regardless of the care and effort taken in gluing up stock to produce large slabs for workbenches, table tops, and the like, the resulting surface invariably will require much exhausting labor to be trued. This is due not to a lack of good workmanship but to the nature of the material. It is virtually impossible to obtain stock that is totally free of even a minimal degree of warp such as crook, bow, cup, and twist. When multiple pieces are joined these normal faults will be quite apparent on the surface.

The workbench top illustrated here is a good example. It was made by face-gluing twenty lengths

The effectiveness of this jig is clearly illustrated in this view in which the jig has been removed from the cutting line.

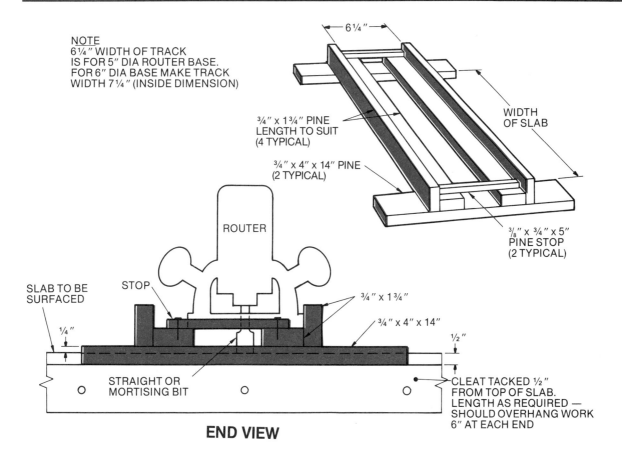

NOTE
6¼" WIDTH OF TRACK
IS FOR 5" DIA ROUTER BASE.
FOR 6" DIA BASE MAKE TRACK
WIDTH 7¼" (INSIDE DIMENSION)

← 6¼" →

WIDTH OF SLAB

¾" x 1¾" PINE
LENGTH TO SUIT
(4 TYPICAL)

¾" x 4" x 14" PINE
(2 TYPICAL)

³⁄₈" x ¾" x 5"
PINE STOP
(2 TYPICAL)

ROUTER

SLAB TO BE
SURFACED

STOP

¾" x 1¾"

¾" x 4" x 14"

¼"

½"

STRAIGHT OR
MORTISING BIT

CLEAT TACKED ½"
FROM TOP OF SLAB.
LENGTH AS REQUIRED —
SHOULD OVERHANG WORK
6" AT EACH END

END VIEW

of stock. The resulting slab turned out as well as could be expected, but there are high and low spots between individual members and an overall concave cupping.

The usual method for truing such a surface would be hand-planing with much follow-up belt-sanding. Quite often the results of this procedure leave much to be desired. But perfection can be achieved with little effort by using the jig shown here with a router and straight cutter. The results will be factorylike. Also, it should be noted that the use of the jig is not limited to new work; it can be used to renew old, worn slabs.

The jig consists of a router track guide that bridges the work. Two ³⁄₄-inch thick boards under the ends of the track ride on two strips that are tack nailed to the sides of the slab. These strips are attached ¹⁄₂ inch below the highest plane of the slab. Both strips must be on a true plane relative to each other. This is critical and should be checked and adjustments made with the aid of a carpenter's level.

When the ³⁄₄-inch-thick track pads are placed on the side strips, the track is suspended at a uniform height above the slab. The router is adjusted for a depth of cut that will just shave wood at the slab's lowest spot.

Starting at one end, the router is passed back and forth across the slab while the track is progressively advanced toward the other end. The width of the track will permit several overlapping passes before it must be advanced to a new position. The results obtained with this jig will undoubtedly make it one of your favorites.

The work will move along at a fast pace if a carbide-tipped cutter is used. Small cross strips at each end of the track serve to stop the router travel at the end of each pass.

Stopped Groove Jig

The stopped groove cut, which begins and ends away from the ends of the workpiece, can be made with a jig consisting of three pieces of wood including a straight fence strip and two wedges cut from a piece of 2×4 stock. The height of the fence strip should be equal to the height of the wedges. The length of the wedges, thus the angle of the incline, is optional: the shorter the wedges, the more abrupt the configuration at the ends of the groove, and vice versa. In any event, the wedges must have sharp ends so the router base will slide down and up the inclines in a smooth, continuous motion.

Jig members are nailed in place with the smallest nails that will hold them so the resulting holes can later be filled without being too conspicuous. This groove was made with a V-groove cutter. Any non-self-piloted cutter can be used.

FENCE

WORKPIECE

WEDGES CUT FROM 2 x 4 STOCK

Clamped Strip Guide

When a narrow edge of work prevents the use of a straightedge guide clamped to the work you can simply attach the guide to the router. Clamp on a strip of wood about two times longer than the diameter of the router base. The long bearing surface will prevent the router from veering off course as frequently happens with standard router accessory guides due to their relatively small bearing surfaces.

Strip is clamped at a measured offset distance from the edge of the bit.

Cutting a groove on a narrow edge is easy and effective with this arrangement.

Jig for Tab-Mounted Frame Back

Frame back panels are frequently installed with minimal regard for neatness. Although the back of the frame is unseen, if fine craftsmanship has gone into the construction of the frame it will be satisfying to treat the hidden part with an equal mark of quality.

With little extra effort and the use of the jig shown, a plywood panel with tabs can be neatly mounted flush with the frame back.

The router with a 1/2-inch-diameter straight cutter is used to make the semicircular mortises. Six strips of wood are used to set up the jig. Two strips are of less thickness than the frame and serve to keep the frame from shifting. The other four strips should be about 1/4 inch thicker than the frame. Two of these are butted against the top and bottom of the frame to control the location of the mortises. The remaining two strips are positioned to stop the travel of the router, thus controlling the length of the mortise.

Back panel is attached to the frame with screws driven through the tabs.

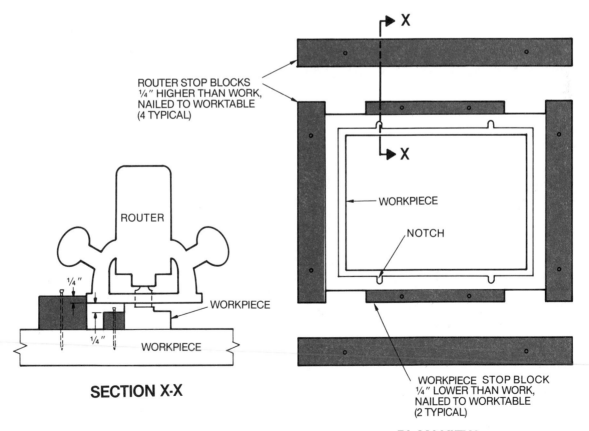

ROUTER STOP BLOCKS 1/4" HIGHER THAN WORK, NAILED TO WORKTABLE (4 TYPICAL)

ROUTER

1/4"

1/4"

WORKPIECE

WORKPIECE

SECTION X-X

X

X

WORKPIECE

NOTCH

WORKPIECE STOP BLOCK 1/4" LOWER THAN WORK, NAILED TO WORKTABLE (2 TYPICAL)

PLAN VIEW

Router is adjusted for a depth of cut equal to the thickness of the back panel. Single pass of the router guided by the strip

Readymade Panel Decorating Jig

If you have neither the time nor the inclination to create your own template guides for decorative cuts in cabinetwork panels you can purchase a readymade jig. The one illustrated here will enable you to rout varying outline designs in work up to 36×36 inches.

With seven sets of corner templates, the jig is easy to set up and use. Four frame rails are attached to the work with special clamps; these, in turn, are secured to the worktable. The templates of the desired design are snapped into the corners of the frame. The router is fitted with a template guide bushing and is simply guided along so the bushing rides against the frame and templates.

A radius-cutting jig is also included in the kit for routing arcs with radii from 5 to $7^1/2$ inches.

These are two of the panel grooving and radial cuts that can be made with the jig.

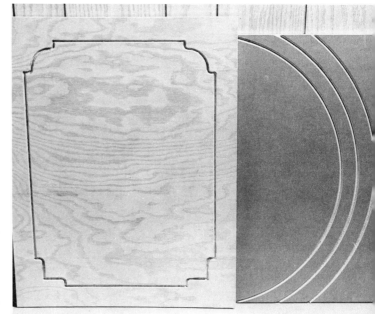

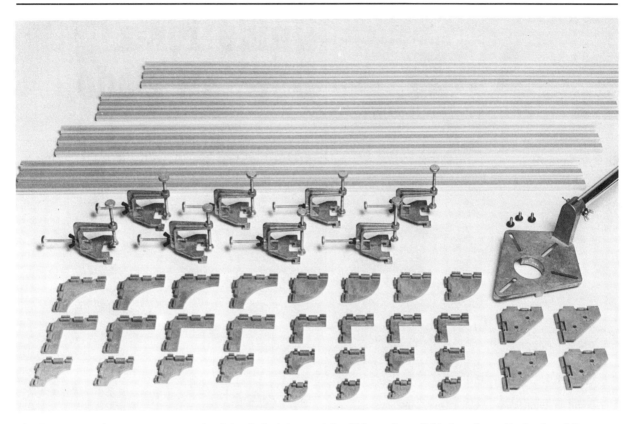

The jig consists of many components, but it is relatively inexpensive. This one is available from Sears, Roebuck and Co.

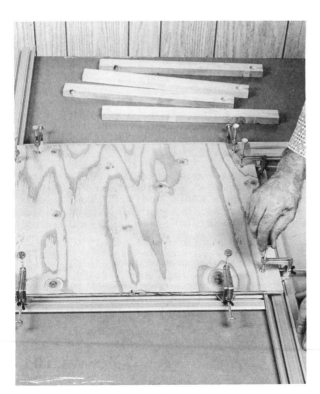

Guide rails are attached with special clamps. Horizontal screws adjust to set the distance from the work edge.

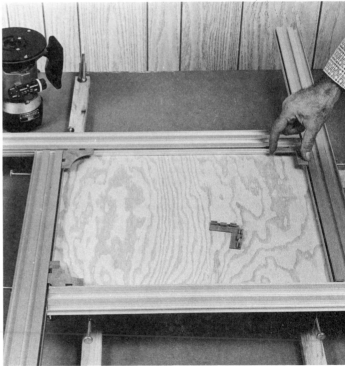

Templates snap into the corners and are interchangeable to produce additional patterns with added passes of the router.

Successful grooving is easy provided the router is kept in constant contact with the guide edges.

The router is attached to a pivoting arm to cut circular grooves in panel.

Guide for a Dovetail Dado

The dovetail dado joint commonly used in drawer construction can be cut with the router using a dovetail bit and the two simple guide setups shown.

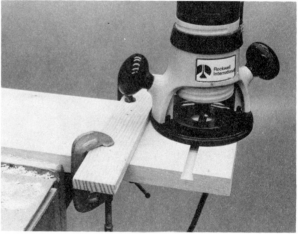

Dado slot is cut in a single pass using a straightedge guide strip clamped to the work.

The dovetail requires two passes with the router. Since the cut is on the end of the stock, a combination router support and guide are set up as shown here. Workpiece is held in the vise at the same height as the base support board. After the first pass has been made the work is rotated in the vise and a second pass is made to complete the dovetail.

Completed joint ready for assembly.

10. LATHE

Semaphore Diameter-Sizing Jig

The semaphore jig is a great aid for duplicating the diameter-sizing parting cuts when multiple turnings of the same design are required. The idea is not new, but the design of this jig is. Other versions of this jig require a somewhat involved initial setup. Not so with this one.

Here is how you work with this jig: the first turning blank is diameter-sized in the usual manner, measuring each groove with calipers. This is then used to adjust the semaphore arms for slip-through. Eliminated is the need for a large number of interchangeable arms of varying lengths or, as is the case with some jigs, the necessity for making minus-factor calculations.

Diameter-sizing cuts on the first turning blank are made with the aid of the calipers in the usual manner. Arms are then adjusted for slip-through. A screwdriver or wrench can be used to lock the lower arms on the shaft.

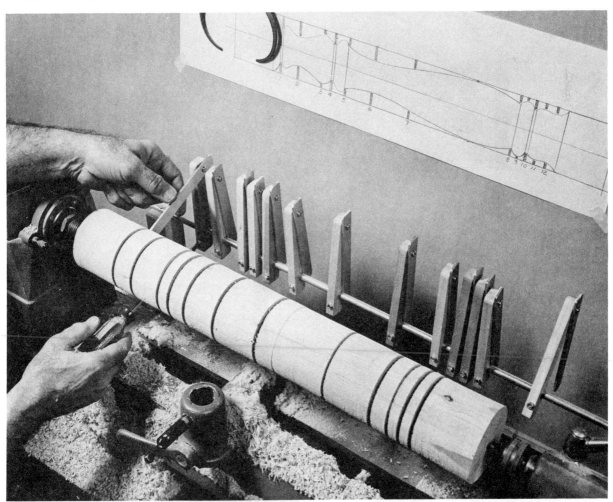

Pre-set arms are rested on the new blank which is then marked with a pencil at the groove locations.

The parting chisel is fed into the work until the arm in line drops through. This is repeated on all subsequent blanks. Each will then have grooves equal in depth to the grooves in the original.

SEMAPHORE DIAMETER-SIZING JIG

MATERIALS LIST

ITEM	QTY	DESCRIPTION
1	1	$3/4'' \times 3\,1/4'' \times 36''$ pine
2	2	$1\,1/2'' \times 1\,1/2'' \times 11''$ pine
3	1	$3/8'' \times 36''$ steel rod
4	2	$1\,1/2''$ # 8 pan-hd screw
5	8	$2''$ finishing nail
6	13	$1/2'' \times 1'' \times 6''$ maple
7	13	$1/8'' \times 3/4'' \times 6''$ aluminum bar
8	13	$1''$ # 8 slotted hex-hd sheet-metal screw
9	13	$1/2''$ # 8 pan-hd screw

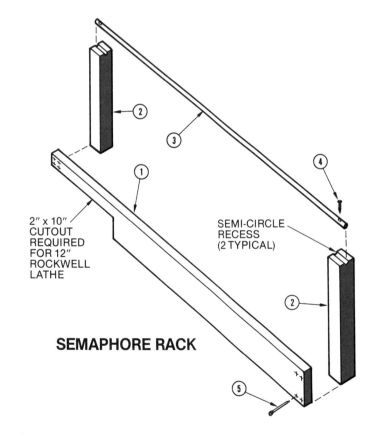

2" x 10" CUTOUT REQUIRED FOR 12" ROCKWELL LATHE

SEMI-CIRCLE RECESS (2 TYPICAL)

SEMAPHORE RACK

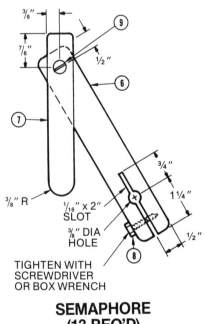

$3/8''$

$7/8''$

$1/2''$

$3/8''$ R

$1/16'' \times 2''$ SLOT

$3/8''$ DIA HOLE

TIGHTEN WITH SCREWDRIVER OR BOX WRENCH

$3/4''$

$1\,1/4''$

$1/2''$

SEMAPHORE (13 REQ'D)

Steadyrest

Make a wraparound steadyrest like this one to keep thin spindles wobble-free without burnishing or burning the wood at the contact point. Friction is avoided because the contact is on three free-spinning ball bearings.

This steadyrest is designed for the Rockwell 12-inch lathe, but the basic concept can be adapted to any lathe. The base-to-center dimension and the method of clamping will likely have to be altered for other lathes.

Plywood can be used throughout with the excep-tion of the roller arms and guide strips; these are made of solid hardwood.

The roller arms must track precisely toward a common center so the guide strips need to be in-stalled in proper alignment. This is easily accom-plished if you proceed as follows: install the bearings and attach the arms with the bolts lightly tightened. Slide the arms forward and into position so the bearings meet at the center. The three will meet only at dead center so measuring or guess-work will not be necessary. Tighten the bolts, then tack nail the strips into place to obtain nail point registration marks. Remove the arms and attach the guide strips with the brads and glue. Check the Yellow Pages under Bearings to find a source of sup-ply for the bearings.

Thin spindle turnings will not flex and wobble when gripped with this efficient steadyrest.

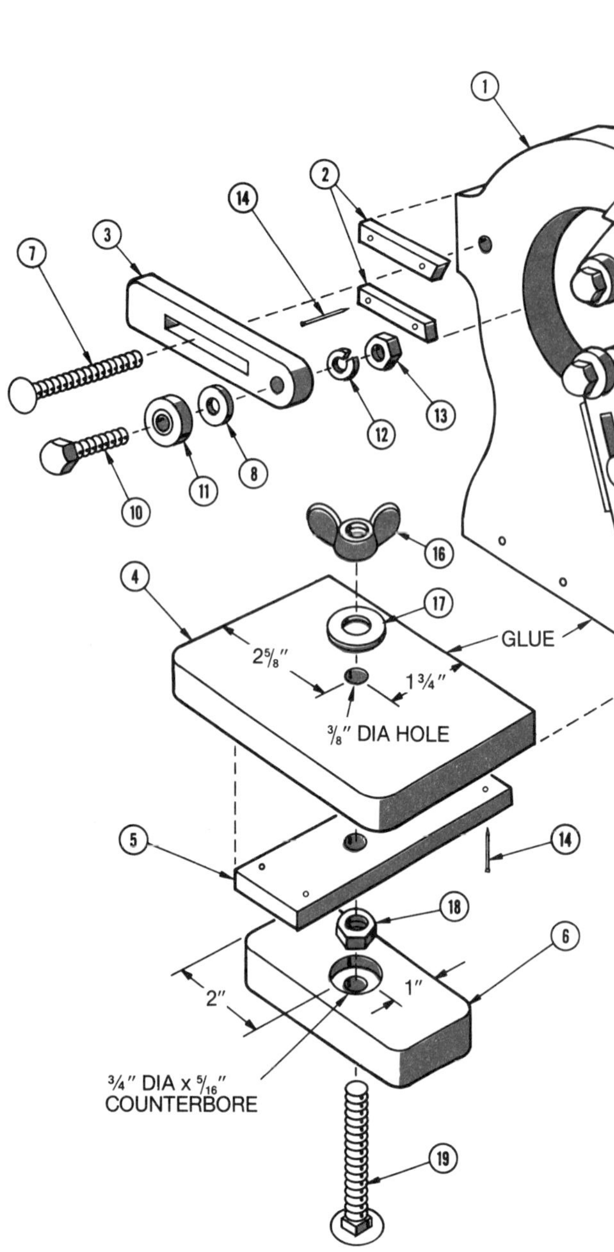
GLUE
2 5/8″
1 3/4″
3/8″ DIA HOLE
2″
1″
3/4″ DIA x 5/16″
COUNTERBORE

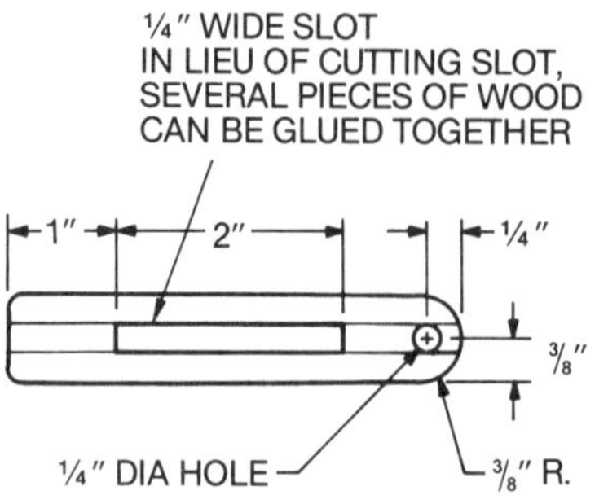
1/4″ WIDE SLOT
IN LIEU OF CUTTING SLOT,
SEVERAL PIECES OF WOOD
CAN BE GLUED TOGETHER
1″
2″
1/4″
3/8″
1/4″ DIA HOLE
3/8″ R.

PART 3

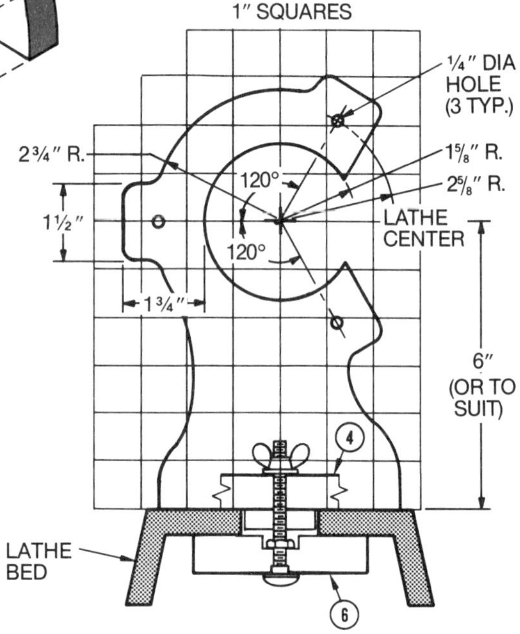
1″ SQUARES
1/4″ DIA
HOLE
(3 TYP.)
2 3/4″ R.
1 5/8″ R.
2 5/8″ R.
120°
120°
1 1/2″
LATHE
CENTER
1 3/4″
6″
(OR TO
SUIT)
LATHE
BED

PART 1

STEADYREST

MATERIALS LIST

ITEM	QTY	DESCRIPTION
1	1	3/4″ × 6″ × 9″ plywood
2	6	5/16″ × 5/16″ × 1 3/4″ hardwood
3	3	7/16″ × 3/4″ × 4″ hardwood
4	1	3/4″ × 3 3/4″ × 5 1/4″ plywood
5	1	3/8″ × 1 1/2″ × 4 1/2″ plywood
6	1	3/4″ × 2″ × 4″ plywood
7	3	1/4″ × 1 1/2″ carriage bolt
8	3	1/4″ flat washer
9	3	1/4″ wing nut
10	3	1/4″ × 1″ hex-hd bolt
11	3	1/4″-thick × 3/4″-dia. × 1/4″-bore ball bearing
12	3	1/4″ lock nut
13	3	1/4″ hex nut
14	16	3/4″ brads
15	3	2″ finishing nails
16	1	3/8″ wing nut
17	1	3/8″ flat washer
18	1	3/8″ hex nut
19	1	3/8″ × 2 1/2″ carriage bolt

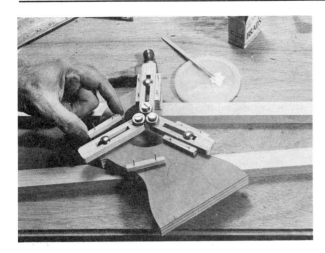

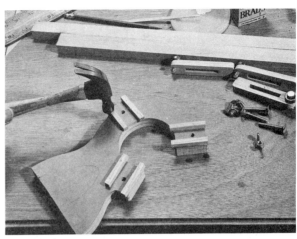

Arms are locked in place when the bearings touch equally at center to facilitate alignment of the guide strips. Arms are pre-slotted by gluing end blocks between the side pieces.

Guides are attached with glue and brads. Overlapping ends are later trimmed flush.

Lathe Duplicator

For projects that involve a number of similar turnings the lathe duplicator will speed the work considerably, and each of the turned pieces will indeed look alike. This duplicator, an accessory for the Rockwell lathe, will justify the investment if your shop activities are heavy on lathe work.

The operation of the duplicator is relatively easy. A half-section hardboard template of the desired spindle shape is mounted on a platform. A stylus on the cutting-tool bar rides against the edge of the template; this causes the cutting tool to duplicate the contours on the turning.

Full-sized pattern is cemented to a piece of hardboard.

Scroll saw is used to cut the template. It will later be split down the middle because a half-pattern only is required.

11. SANDERS

Sanding Aids for Routed Work

fanned pattern in size order. Place the smallest in front, then wrap some tape around the end. Chuck the bundle in the drill press at a slight downward angle so some of the flaps will strike the bottom corner in the work. The result will be a smooth edge blending with the already sanded flat surface.

Among the numerous sanding accessories that are commercially available, none are suitable for a rather common shop requirement—smoothing the surfaces of routed recesses. Consequently, tedious hand-sanding is relied on to eradicate tool marks and to smooth surfaces to finish-ready condition. But there is an easier way.

You can make a bottom sanding drum to level and smooth the flat surface of the recess and use a bundle of abrasive strips to reach curved inside corners.

To make the drum, insert a hanger bolt into a small wood disc of a diameter suited to the particular project. Cut a piece of abrasive paper about 3/4 inch larger in diameter than the diameter of the disc. Outline a circle the diameter of the disc on the back of the abrasive paper, then use a scissors to make closely spaced radial cuts up to the line. Use rubber cement to attach the abrasive piece to the disc. Fold the cut segments upward so they flare slightly out. When the disc is rotated at high speed in a drill press, the turned-up sides of the abrasive will fan out slightly; this prevents a sharp demarcation at the lower corners as the bottom of the disc sands the flat surface of the work.

The flexible abrasive bundle is made by cutting strips of varying widths and lengths ranging in width from 1/8 to 3/8 inch and in length from 1 to 3 inches. Make one or two small cuts into one end of each strip. Stack the ten or twelve strips in a slightly

The flaps work their way into the curved edges, smoothing them evenly all the way around.

Strips are arranged in a fanned random pattern with the smaller strips low and at the front.

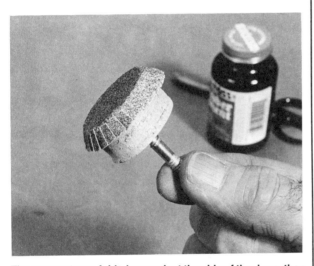

Cut closely spaced segments in the sandpaper up to the diameter circle of the drum.

The segments are folded up against the side of the drum; they will angle out slightly. Rubber cement is used to attach the abrasive disc so it can be removed easily for replacement.

Smoothing a routed recess the easy way.

Scroll Saw Sanding Block

The usual sanding accessory available for the scroll saw is rather narrow, thus making it somewhat difficult to sand broad surfaces flat and true. Better work can be done with a homemade oversize sanding block.

Insert a hanger bolt into the end of a wood block and chuck the projecting end into the saw. Use rubber cement to attach the abrasive sheet to the face of the block. For irregular edges you can shape the block to suit.

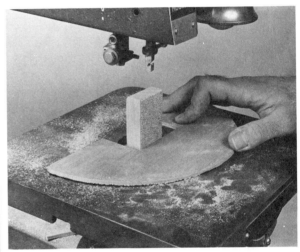

The broad contact surface of this sanding block is not apt to dig in erratically; this makes it easy to sand edges true and flat. The block also does well on sharp inside corners.

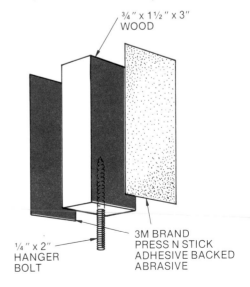

¾" x 1½" x 3" WOOD

3M BRAND PRESS N STICK ADHESIVE BACKED ABRASIVE

¼" x 2" HANGER BOLT

Round Stock Tapering Jig

The ends of dowels and other rounds can be tapered accurately by feeding them against a guide strip clamped to the disc-sander table at the desired angle. A board is clamped to the overhanging section to form a ledge on which the stock rests. If more than one piece is to be shaped, a stop can be attached to the board to block the lateral movement, thus ensuring uniformity.

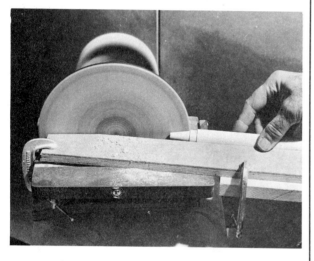

Stock is rotated fully before each forward movement toward the center of the disc. The jig can be used for specialty cuts as well. In this example the guide strip was first clamped parallel to form the straight shoulder.

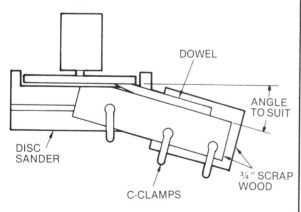

Circular Sanding Jig

Circular workpieces cut with a band, scroll, or sabre saw must be sanded to remove the saw ripples and to achieve a finish-ready edge. The stationary disc or belt sander is the tool best suited for this purpose. However, if you do not own such equipment you can nevertheless do a superb job of sanding the edges by using a portable finishing sander with this simple jig.

The sander holding block illustrated is made for a specific sander, the Rockwell Model 4485, but you can alter the configuration of the cutout to suit your particular sander.

To sand a disc, insert it between the handles, take up lightly on the nut and bolt axle, then press down lightly on the handles while slowly rotating the disc. If you prefer to use a smaller pivoting hole in the work, substitute a nail for the bolt.

Two More Jigs for Sanding Discs

Here are two variations on the same theme of utilizing the finishing sander in a stationary capacity for sanding circular discs. With both methods the sander is held in position by makeshift clamping methods. Notice that regardless of the physical shape of the sander it must be propped on a board with the moving pad section overhanging the edge. This provides clearance to allow the pad to move freely.

Smaller work does not need the 1 x 2 side supports; it can be handled with a wide tiller only. Notice how the flat-sided handle of this particular sander lends itself to positive clamping.

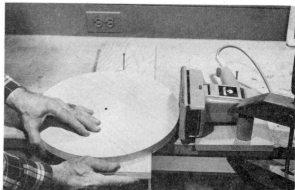

Portable finishing sander is held captive in the block to function as a stationary sander. Cloth-backed abrasive is suggested because it is tougher and longer lasting.

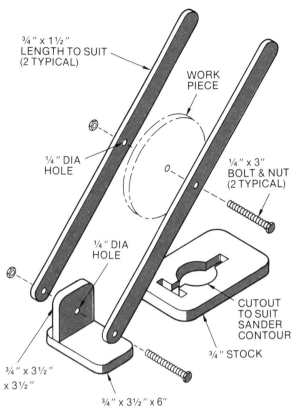

¾" x 1½"
LENGTH TO SUIT
(2 TYPICAL)

WORK
PIECE

¼" DIA
HOLE

¼" x 3
BOLT & NUT
(2 TYPICAL)

¼" DIA
HOLE

CUTOUT
TO SUIT
SANDER
CONTOUR

¾" STOCK

¾" x 3½"
x 3½"

¾" x 3½" x 6"

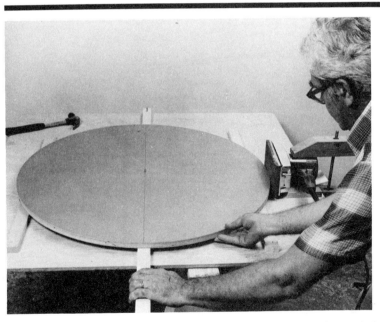

Edges of circular work of unlimited diameter can be sanded with this arrangement. Sander is clamped to the table with the moving area overhanging a piece of ¾-inch stock. Wedge is inserted under the slanted housing to plumb the sanding pad perpendicular to the worktable. Two strips of 1 x 2 near the outside edges of the work serve to keep it level. Center strip serves as a tiller; nail through the far end permits work to be swung into contact with the sander. Nail through the center of the work allows it to be rotated in a true path. Light pressure is applied to the tiller while the work is rotated.

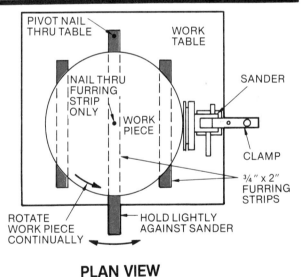

PIVOT NAIL
THRU TABLE

WORK
TABLE

NAIL THRU
FURRING
STRIP
ONLY

WORK
PIECE

SANDER

CLAMP

¾" x 2"
FURRING
STRIPS

ROTATE
WORK PIECE
CONTINUALLY

HOLD LIGHTLY
AGAINST SANDER

PLAN VIEW

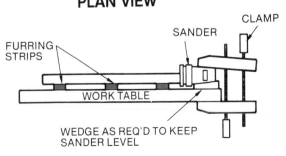

CLAMP

SANDER

FURRING
STRIPS

WORK TABLE

WEDGE AS REQ'D TO KEEP
SANDER LEVEL

END VIEW

Sanding Jigs for Lathe

The lathe need not stand idle between turning projects. It can instead be put to use for sanding chores with the jigs shown here. The vertical backboard jig is used to guide work against a sanding drum of 3-inch or smaller diameter.

The pipe and flange supported horizontal table can be used with both the drum- and disc-sanding accessories. The lathe's tool rest clamp holds this unit firmly in place while permitting vertical and horizontal adjustments.

Rear view shows how the jig is secured to the lathe bed. Ribbed design gives the unit necessary rigidity.

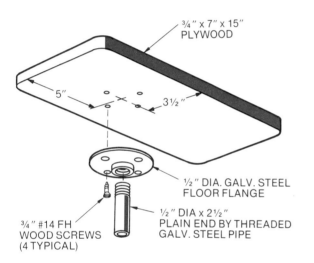

¾" x 7" x 15" PLYWOOD

5"

3½"

½" DIA. GALV. STEEL FLOOR FLANGE

¾" #14 FH WOOD SCREWS (4 TYPICAL)

½" DIA x 2½" PLAIN END BY THREADED GALV. STEEL PIPE

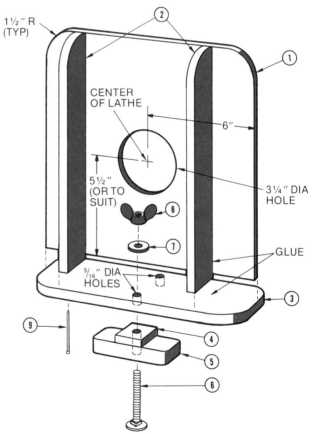

1½" R (TYP)

CENTER OF LATHE

6"

5½" (OR TO SUIT)

3¼" DIA HOLE

GLUE

⁵⁄₁₆" DIA HOLES

VERTICAL BACKBOARD

MATERIALS LIST

ITEM	QTY	DESCRIPTION
1	1	½" × 12" × 12" plywood
2	2	½" × 2¼" × 12" plywood
3	1	½" × 5" × 12" plywood
4	2	⅜" × 1⅜" × 2" plywood
5	2	¾" × 2" × 4⅜" plywood
6	2	⁵⁄₁₆" × 2½" carriage bolt
7	2	⁵⁄₁₆" flat washer
8	2	⁵⁄₁₆" wing nut
9	16	1½" finishing nail

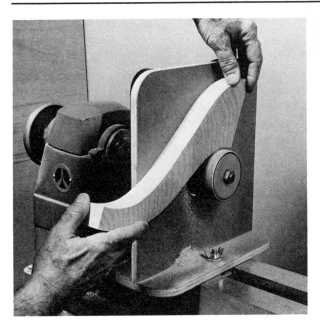

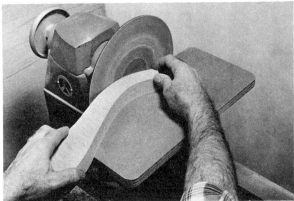

Curved edges can be sanded perfectly true with this arrangement. The work is fed *against* the rotation.

The sanding disc is used for outside corners.

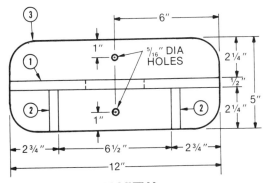

PLAN VIEW

3

1

5/16" DIA
HOLES

6"

1"

2¼"

½"

2¼"

1"

5"

2¾"

6½"

2¾"

12"

2

2

An otherwise difficult task made easy: shaping a uniform concave curve on a leg such as this so it will mate with a cylindri-
cal post. A strip temporarily nailed to the leg rides true against
the clamped guide strip nearest the drum. Strip on the other
edge of the platform keeps the work on a true horizontal plane.
The work is moved back and forth against the sanding drum
until the guide strips meet. Drum diameter should match that
of the post or vice versa.

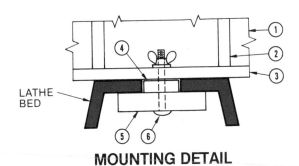

LATHE
BED

1

2

3

4

5   6

MOUNTING DETAIL

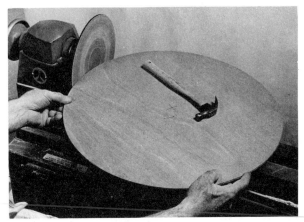

A circular edge is handled in this manner: the work is pressed
firmly against the disc and a pivot nail is then driven through
the work center and into the platform. Work is rotated clockwise after the power has been turned on.

Belt Sander Conversion Jig

Although the portable belt sander excels for stock removal and smoothing tasks on large workpieces, it is not suitable for doing controlled precision sanding on small work. The latter is best done on a stationary belt sander.

Lacking a stationary belt sander, you can convert your portable into a stationary model with this jig. The sander mounts in a matter of seconds and can be removed just as quickly for regular use. To make the conversion, you have only to place the sander in the cradle, insert the holding bar, and tighten two thumbscrews. The tilting table permits precise bevel sanding.

The cradle is sized and shaped to accept the Rockwell Model 337—3-by-21-inch belt sander and several others of the same make with similar housings. Other sanders can be accommodated by making minor changes. If it should be necessary to alter the configuration of the cradle, it is advisable to make a cardboard template of the sander's housing outline. Get a perfect fit by trial and error on cardboard before you transfer the outline to the wood.

The holes in the cradle members are purposely made slightly oversize to allow clearance for the dowel hold-down to bear down freely on the sander's handle.

Any wood can be used throughout, but the pressure pads that bear on the dowel ends should be made of hardwood. Two partly driven screws and slightly oversized holes give the pads the necessary free movement.

MATERIALS LIST

ITEM	QTY	DESCRIPTION
1	1	3/4" × 83/8" × 14" pine
2	1	3/4" × 73/4" × 12" pine
3	1	3/4" × 71/4" × 12" pine
4	2	3/4" × 7" × 71/4" pine
5	2	3/4" × 2" × 4" pine
6	2	1/2" × 1" × 13/4" pine
7	1	3/4" × 53/8" × 12" pine
8	2	3/4" × 11/2" × 8" pine
9	2	13/16" × 13/16" × 4" oak or maple
10	1	3/4" × 12" dowel
11	2	3/8" × 1" dowel
12	1	11/2" × 12" strip hinge (with 3/4" # 6 fh screws)
13	2	6" lid support (Brainerd # 1085XC or equiv.)— remove friction fitting
14	2	3/4" screw eye
15	4	1/4" flat washer
16	2	1/4" × 11/2" thumbscrew
17	2	1/4" tee nut
18	4	11/2" # 6 rh wood screw
19	16	2" finishing nail
20	8	2" common nail
21	4	3/16" flat washer

Using the tilt table to true a miter.

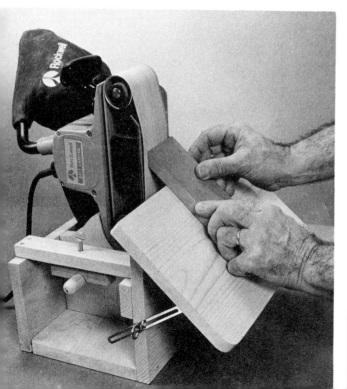

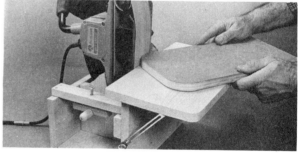

Fixed belt sander makes easy work of precision sanding.

The dowel hold-down is tensioned by turning two thumbscrews. This bears the dowel against the handle and holds the sander firmly in place.

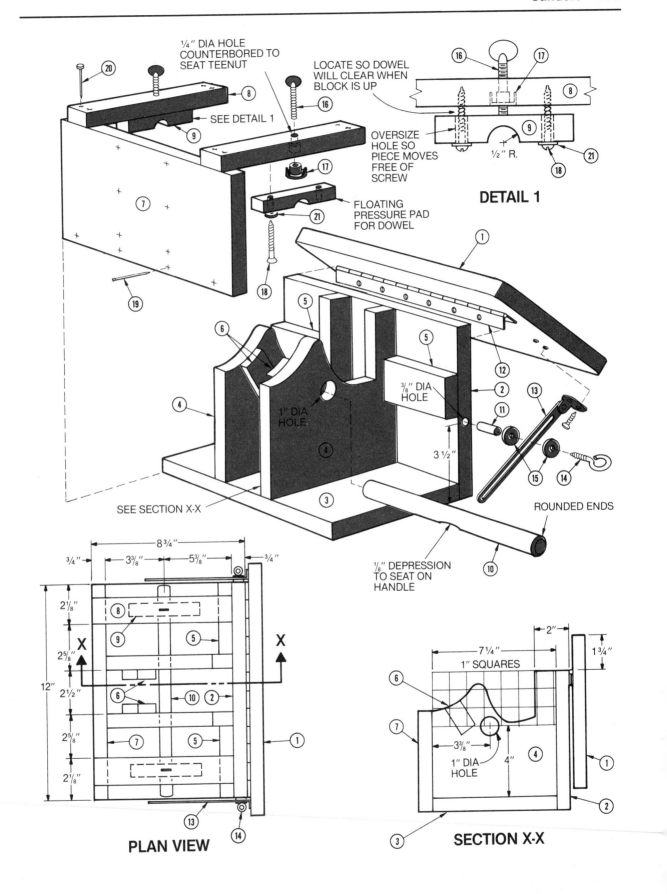

¼″ DIA HOLE
COUNTERBORED TO
SEAT TEENUT

LOCATE SO DOWEL
WILL CLEAR WHEN
BLOCK IS UP

SEE DETAIL 1

OVERSIZE
HOLE SO
PIECE MOVES
FREE OF
SCREW

½″ R.

DETAIL 1

FLOATING
PRESSURE PAD
FOR DOWEL

1″ DIA
HOLE

3/8″ DIA
HOLE

3½″

1″ DIA
HOLE

3 ½″

SEE SECTION X-X

ROUNDED ENDS

1/8″ DEPRESSION
TO SEAT ON
HANDLE

8¾″

¾″   3⅜″   5⅜″   ¾″

2⅛″

X   X

2⅝″

12″

2½″

2⅝″

2⅛″

PLAN VIEW

2″

7¼″

1″ SQUARES

3⅜″

1″ DIA
HOLE

4″

1¾″

SECTION X-X

12. ASSEMBLY

Cross Bars for Edge - Gluing

After glue has been applied in the process of edge-gluing boards to produce a slab, the first thing that needs to be done is to clamp cross supports lightly at several locations. They serve to keep the slab from cupping when the bar clamps are applied to the edges.

There will be no need to scrounge for scrap wood or continually mar good wood for the purpose if you make a few of these reuseable, self-clamping, adjustable cross bars.

Make them of hardwood—ash, maple, or oak. Glue three spacers between each pair of strips to form the slots for the carriage bolts. In order to prevent glue drippings from sticking the bars to the work, apply several coats of clear satin finish; then apply a coating of paste wax.

Bars are only snugged up to the work—never tightened so much as to prevent the bar clamps from closing the joint. This beats fumbling with C clamps and separate pieces of wood.

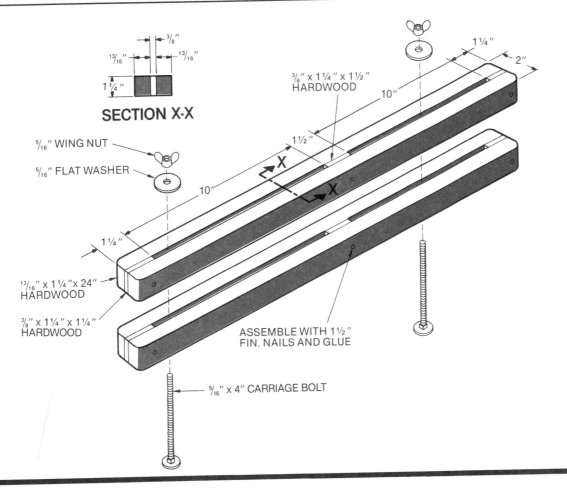

SECTION X-X

$\frac{5}{16}$" WING NUT

$\frac{5}{16}$" FLAT WASHER

$\frac{3}{8}$" x 1 $\frac{1}{4}$" x 1 $\frac{1}{2}$"
HARDWOOD

$\frac{13}{16}$" x 1 $\frac{1}{4}$" x 24"
HARDWOOD

$\frac{3}{8}$" x 1 $\frac{1}{4}$" x 1 $\frac{1}{4}$"
HARDWOOD

ASSEMBLE WITH 1 $\frac{1}{2}$"
FIN. NAILS AND GLUE

$\frac{5}{16}$" x 4" CARRIAGE BOLT

Clamping Pad for Dovetails

A dovetail joint is usually made with the pins and tails projecting slightly above the face of the work. They are then sanded flush. The projections prevent the use of clamp pads and clamps directly over the joint. To do so, cut pin clearance notches in the clamp pad. The pad can then be placed directly on the dovetails.

The notched clamp pad misses the projecting pins, directing the pressure squarely on the dovetail. Pins are self-tightening and need not be clamped.

Frame Miter Clamping Jig

This frame gluing jig applies diagonal pressure evenly to the four corners at the same time to close the miter joints tight and true. The single clamp on this jig applies uniform pressure at the four corners. The corner blocks should not be firmly tightened; they should be free to pivot.

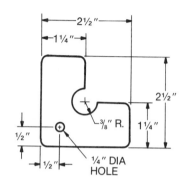

DETAIL 1

MATERIALS LIST

ITEM	QTY	DESCRIPTION
1	4	$^{13}/_{16}" \times 2" \times 18"$ hardwood
2	2	$^{13}/_{16}" \times 3" \times 5"$ hardwood
3	4	$^{13}/_{16}" \times 2^{1}/_{2}" \times 2^{1}/_{2}"$ hardwood
4	8	$^{1}/_{4}" \times 2"$ fh bolt
5	8	$^{1}/_{4}"$ flat washer
6	8	$^{1}/_{4}"$ wing nut

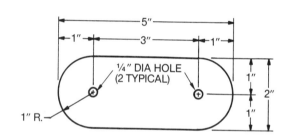

DETAIL 2

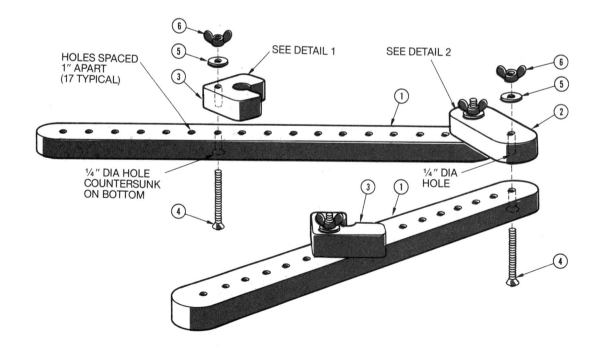

Corner blocks can be located in any set of holes in the arms to permit clamping frames of varying sizes.

Corner Clamp Pads

Simplify the clamping of table legs into a corner with the aid of V-notched clamp pads cut to match the outside angle. Applied diagonally in the manner shown, the clamp pressure is advantageously directed evenly into the corner.

Right-angle notches are cut at 45 degrees to the surface to make the corner clamp pads.

13. METALWORK

Curlicue Former

The curled ends of this table frame are made by hand-bending ⅛-by-1-inch hot-rolled mild steel stock over a two-part jig. It should be noted that cold drawn stock of this size is too tough to be formed with the procedure outlined here.

The key part of the jig is made with a piece of ¾-inch stock sandwiched between two pieces of ⅛-inch thick aluminum. The mouth of this piece is sized to receive the work stock with a snug, slightly forced fit to prevent it from sliding out during bending. As with almost any metal-bending operations, all the bend lines should be marked before the first

A simple bending jig was used to form the metal frame for this plywood slab table.

End of the bar is inserted into the slot to make the bend. Once the bend is started the piece tends to lock itself in place. Lifting straight up will free it.

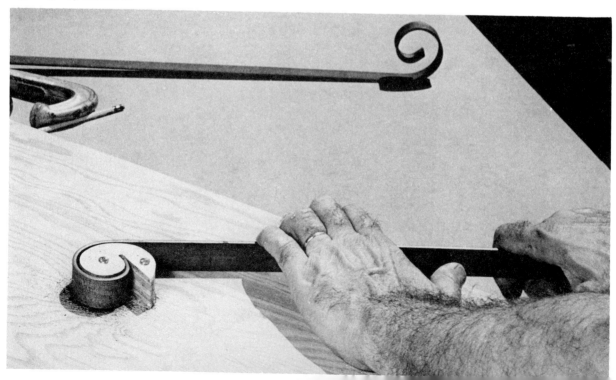

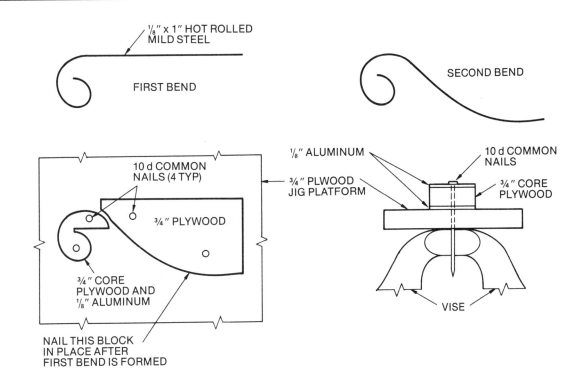

⅛" x 1" HOT ROLLED MILD STEEL

FIRST BEND

SECOND BEND

10 d COMMON NAILS (4 TYP)

¾" PLYWOOD

¾" CORE PLYWOOD AND ⅛" ALUMINUM

NAIL THIS BLOCK IN PLACE AFTER FIRST BEND IS FORMED

⅛" ALUMINUM

¾" PLWOOD JIG PLATFORM

10 d COMMON NAILS

¾" CORE PLYWOOD

VISE

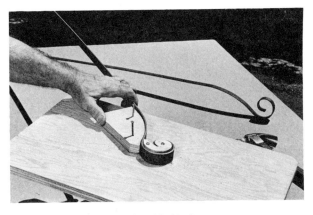

Second section of the jig is added to form reverse curve.

A block of wood held close to the vise jaw is used to form the small-radius corner bend.

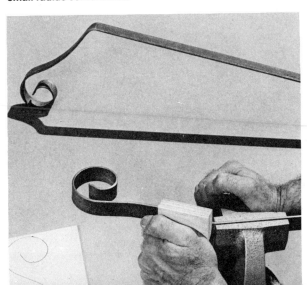

bend is made. Any attempt to make a lengthwise measurement over a bent surface will likely yield erratic results.

In the first stage, the end of the stock is inserted into the mouth of the forming block and pulled around to a predetermined spot marked on the backboard. When all the pieces have been so bent, the second part of the jig is installed. This is a plywood block shaped to form the reverse curve. It is tack nailed into place, and the end-bent workpieces are reinstated in the mouth to continue the forming. The corner bends are made by securing the workpiece in the jaws of a vise and using a block of wood to bear firmly against it.

The frame is assembled with rivets. They are peened over a steel block.

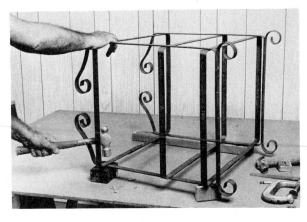

Forming Heavy Stock

The designation "heavy," as used here, is relative. The ³/₁₆-by-1¹/₂-inch hot-rolled mild steel stock used in the project illustrated would be considered light for certain applications; however, it is about the limit for forming with the jigs shown in this chapter.

The oak and steel bench shown here features framing with a variety of bends that are made with several simple jigs. The toughest bend of all to make is the small-radius corner near the end of the stock, but this is readily accomplished with a jig (#1) made with a door hinge mounted back side up between two lengths of 2 × 4.

Selection of a tough hinge for this purpose is very important. Bargain-priced, off-brand hinges usually are made of poor-quality steel and will fracture behind the knuckles when subjected to the stress encountered in this application. A Stanley 3¹/₂-by-3¹/₂-inch No. 758 hinge holds up well for this use.

The two outer holes in each leaf are rebored to permit inserting ⁵/₁₆-inch bolts. The two-knuckle leaf is attached to the shorter length of 2 × 4.

The jig (#2) for forming the sweeping seat curves is cut from 2 × 4 stock. The forming jig (#3) for the arm rest is cut from 4/4 hardwood. The plan is provided to serve as a guide. You can adapt the basics

This bench reflects the potential of combining wood and metal fabrication.

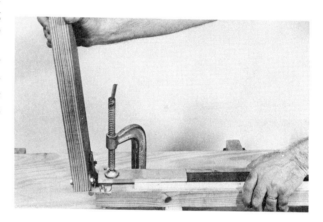

Strip to be bent is clamped between the hinge and a steel hold-down. Wood strip near the forward edge aligns the workpiece parallel to the jig.

Small-radius bending jig is made by attaching a hinge, back side up, between two lengths of 2 × 4. Hinge is let into a rabbet on the stationary member so it lies flush with the surface to accommodate the heavy stock.

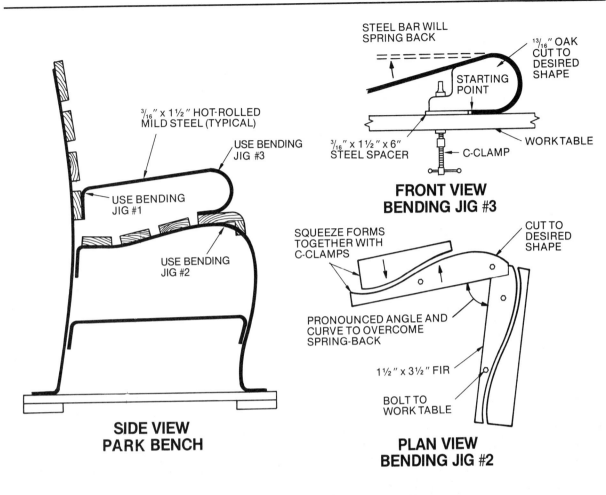

STEEL BAR WILL
SPRING BACK

$^{13}/_{16}$" OAK
CUT TO
DESIRED
SHAPE

STARTING
POINT

$^3/_{16}$" x 1½" x 6"
STEEL SPACER

C-CLAMP

WORK TABLE

**FRONT VIEW
BENDING JIG #3**

$^3/_{16}$" x 1½" HOT-ROLLED
MILD STEEL (TYPICAL)

USE BENDING
JIG #3

USE BENDING
JIG #1

USE BENDING
JIG #2

**SIDE VIEW
PARK BENCH**

SQUEEZE FORMS
TOGETHER WITH
C-CLAMPS

CUT TO
DESIRED
SHAPE

PRONOUNCED ANGLE AND
CURVE TO OVERCOME
SPRING-BACK

1½" x 3½" FIR

BOLT TO
WORK TABLE

**PLAN VIEW
BENDING JIG #2**

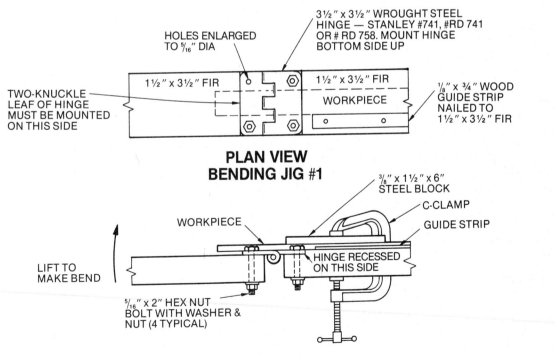

HOLES ENLARGED
TO $^5/_{16}$" DIA

3½" x 3½" WROUGHT STEEL
HINGE — STANLEY #741, #RD 741
OR # RD 758. MOUNT HINGE
BOTTOM SIDE UP

1½" x 3½" FIR

1½" x 3½" FIR

WORKPIECE

$^1/_8$" x ¾" WOOD
GUIDE STRIP
NAILED TO
1½" x 3½" FIR

TWO-KNUCKLE
LEAF OF HINGE
MUST BE MOUNTED
ON THIS SIDE

**PLAN VIEW
BENDING JIG #1**

$^3/_8$" x 1½" x 6"
STEEL BLOCK

C-CLAMP

GUIDE STRIP

WORKPIECE

HINGE RECESSED
ON THIS SIDE

LIFT TO
MAKE BEND

$^5/_{16}$" x 2" HEX NUT
BOLT WITH WASHER &
NUT (4 TYPICAL)

**FRONT VIEW
BENDING JIG #1**

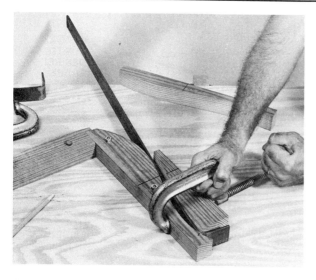

Front leg contour is formed by squeezing the mating curved blocks together.

for various applications. The procedure in using the jigs follows.

A strip is cut to length for the front leg/seat section. A slight radius is ground on the end to prevent snagging on the wood. The bend line is marked on the strip. The mark is centered over the hinge and clamped to the jig together with a hold-down—a piece of steel $3/8$ by $1^1/2$ by 5 inches. The clamp is applied close to the hinge, and the short leg of the jig is raised to form the bend.

The strip is then placed against the second jig and formed by clamping on the mating block. With the clamps still in place, the remainder of the strip is bent around the corner by hand. The seat curve is not yet made. The mark for the second right-angle bend is made by tracing against the end of the jig end.

The strip is taken back to the hinged jig to make

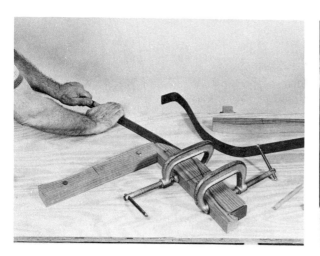

The strip is brought around the corner by hand. Edge of the strip is kept in contact with the worktable.

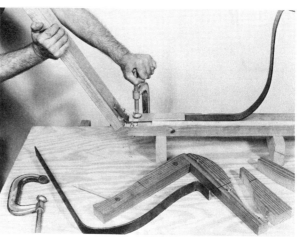

Making the second corner bend. This illustrates why the seat contour bend is made later. If it had been made in advance, the small-radius bend would not be possible because the curved section would prevent clamping in the jig.

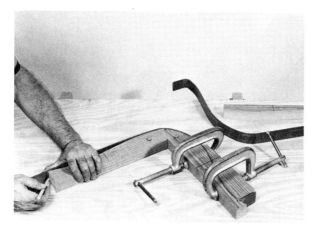

Strip is held against the end of the jig and marked for the small-radius end bend. Work is temporarily removed after the mark is made and transferred to the first jig.

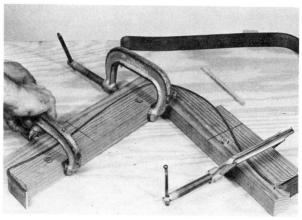

Work is returned to the contour jig to make the seat bend.

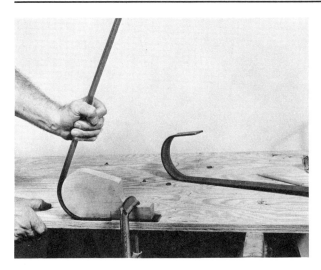

Armrest forming jig is clamped vertically to the worktable. Scrap strip under the clamp provides a gap for inserting work.

the second end right-angle bend. The piece is again moved to the second jig to make the seat curve. Notice that the fixed sections of this jig are positioned at an exaggerated angle to each other to allow for normal spring-back. This factor must always be considered when forming metal. The amount of spring-back is a variable and is best determined by a preliminary test.

The back upright is bent freehand because a jig of the appropriate size would simply waste lumber. The strip is placed over two sawhorses and topped with a strip of 1/2-by-2-inch hardwood. Downward hand pressure is applied to form the shallow curve.

The armrest is formed by hand-bending over the shaped block. The block is clamped vertically to the workbench over a piece of scrap strip. The end of the strip to be bent is placed in the gap as shown.

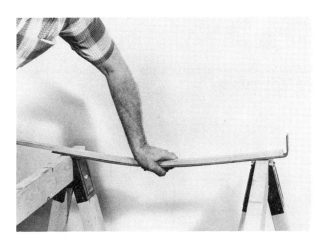

Strip of hardwood and two sawhorses are used to form the shallow curve for the back member.

Oak slats are marked for drilling. The final assembly is done after painting the metal and varnishing the wood.

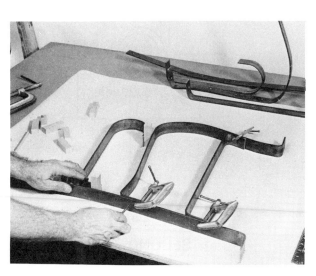

Parts are assembled with clamps to mark bolt hole locations.

A clamp is placed over the high point of the block to hold it down during the initial stage of the bend to prevent it from lifting. When the strip reaches the clamp, the clamp is removed and the bend is continued. The piece is placed on the hinged jig to make the final right-angle bend.

As you can see, these jigs are easy to make and use. You need only change the dimensions and shapes of curve-bending jigs to meet practically any requirement.

Small-Radius Bends in Aluminum

Aluminum bar stock up to ¼ by 1 inch can be formed into loops of small radii with the aid of a shaped hardwood block and a tee pipe fitting. A curved cutout is made in the block equal in radius to the radius of the pipe fitting plus the thickness of the bar stock to be bent. A second, smaller block with a shallower cutout is also required. The corners of the cutouts are rounded in order to give the stock a start in bending. The tee fitting has a thicker, raised section at the end. This is ground flat on two sides of the center to allow the work to make full flat contact along the sides.

The work is done in a metalworking vise using the tee fitting as the fulcrum. The bends are made in stages as shown.

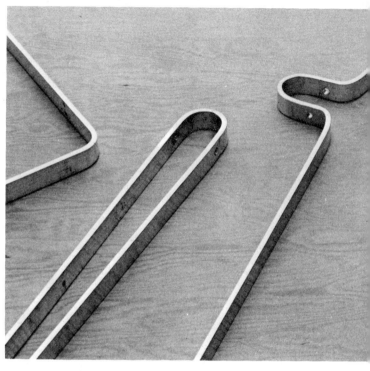

Typical loop bends that can be made with the block and tee fitting jig shown below.

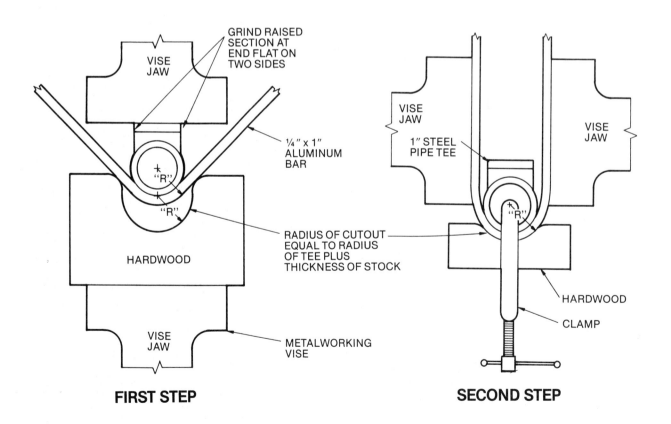

FIRST STEP

SECOND STEP

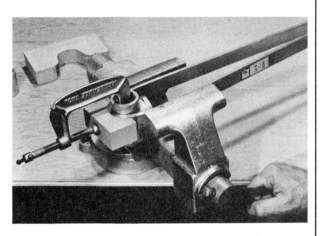

The bar is sandwiched between the block and a 1-inch tee pipe fitting and pressed into a bend until the vise jaw interferes. Work and block are now removed from the vise.

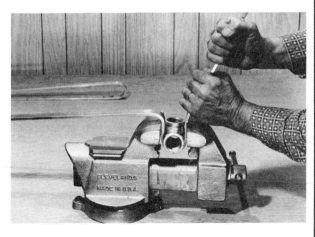

The smaller block is clamped, together with the tee, to the bent section of the bar. The bar is bent freehand until it can fit into the vise jaws which are opened wide. The vise is then closed until the sides of the bar touch the sides of the fitting to form a U.

Return bends are made by hand by gripping the bar close to the bend at the vise.

Light-Duty Scrolling Jig

Mild steel bar stock, $1/16$ by $1/2$ inch is used for the scrolled arms of this lathe-turned sconce. The former block is cut from a piece of $1/2$-inch hardwood that is then glued to a back-up block. Due to the sweep of the curve, which continues into the slot, a slight bend is pre-formed on the lead end of the bar to permit it to fit into the curved slot. This is done freehand by inserting about $1/8$ inch of the bar in a vise, then bending slightly.

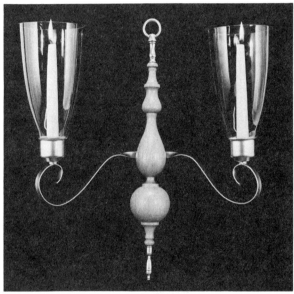

Graceful metal arms add a nice touch to this turned spindle. The right-angle twist in the arms is made by clamping the strip in a vise and using a wrench to make the turn.

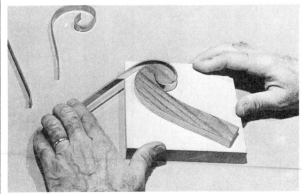

The hardwood former must be glued to the baseboard to prevent it from splitting.

Bending Jig for Aluminum Tubing

Thin-wall aluminum tubing can be bent successfully with simple homemade jigs. Care must be exercised, however, in observing minimum radii recommendations to avoid crimping. Round tubing for home shop use is commonly available in diameters of ³/₄, ⁷/₈, 1, and 1¹/₄ inches. Respectively, the minimum bending radii are 5, 7, 9, and 11 inches.

The jig shown here is used for making 5-inch radius bends in ³/₄-inch tubing. It consists of a 10-inch-diameter disc and straight supporting blocks. The inside edges of the inside blocks must be positioned tangent to the disc. Two screws are used to hold the blocks to the backboard, and they should be installed near the ends so the center area will be free to flex slightly under clamp pressure in order to hold the tubing firmly in place. The blocks are spaced so the tubing fits snugly between them.

A single set of support blocks will do for making bends on one plane, but if the project involves bends in two planes a second set of support blocks (as shown) will permit making right- and left-hand bends on the second plane.

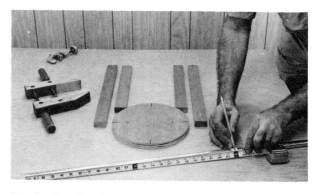

Circular bending form is index-marked at right-angle positions. All bend start marks are made on masking tape for good visibility. Marks on tube are aligned with the marks on the form to assure proper curve.

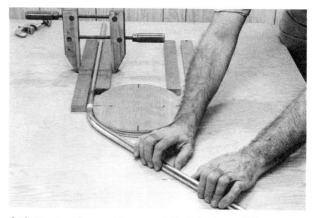

A clamp must be used to prevent the tube from sliding during bending. If the tube slides, the bend will be mislocated. Also, sliding could result in crimping. The tube must be pulled around the form slowly and steadily.

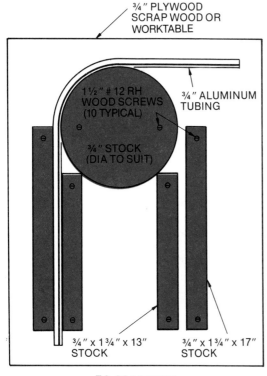

³/₄″ PLYWOOD
SCRAP WOOD OR
WORKTABLE

1½″ # 12 RH
WOOD SCREWS
(10 TYPICAL)

³/₄″ ALUMINUM
TUBING

³/₄″ STOCK
(DIA TO SUIT)

³/₄″ x 1³/₄″ x 13″
STOCK

³/₄″ x 1³/₄″ x 17″
STOCK

PLAN VIEW

Jig should be set up parallel to the table edge so that right-angle bends can readily be checked with a large square.

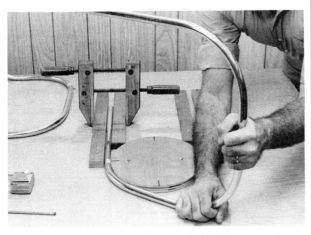

Notice how the last bend on this piece is made out of sequence. This was done in order to avoid a "trap." If the bends were made in consecutive sequence, the last bend would have met with interference from the jig. The bending sequence should always be planned in advance.

This shows why the second set of straight supports are needed for making bends on a second plane. To match this bend on the section above, the tube will have to be placed in the opposite supports.

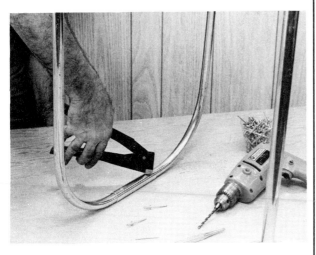

A Pop Riveter is a must for securing sleeved ends.

Clean Cuts in Thin Stock

Cutting thin metal stock clean and true can be difficult when the usual methods and tools are used. Here are a few tricks that will enable you to obtain good results when working with relatively soft nonferrous metals such as aluminum, brass, and copper.

Tape the metal sheet to a slightly oversize backboard of ¼-inch plywood. A panel of the same thickness is tack nailed above, beyond the metal, to form a sandwich. Notched push stick is used to bear down on the pack and safely push it over a table saw blade. A fine-tooth plywood blade will make a clean cut in light-gauge aluminum, copper, or brass sheet.

Sharp utility knife and clamped metal straightedge are used to cut thin aluminum sheet clean and true.

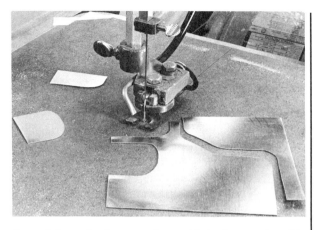

Down-deformed edges can be avoided when cutting thin metal stock on the scroll saw by using a zero-gap table insert. Cut half way into a piece of ⅛-inch hardboard about the size of the scroll saw table. Tape this to the table, then cut the metal as desired. It will not deform because there will be no gap between the hardboard and the blade.

To make a radial cut on the end of metal tubing, insert it into a hole of the same diameter in a block of wood. Draw the desired curve on the block and make the cut.

The procedure is ideal for making T joints and can be used for wood stock as well.

Bending Jig for Thin Sheet Metal

Sharp corner bends in light-gauge sheet metal can be made accurately with a back flap jig. The jig is made to the size required with two pieces of ¾-inch plywood and a length of continuous hinge.

The hinge is set into rabbeted recesses so that it lies flush with the surface of the wood. The hinge is installed back side up so it will form a sharp-angle corner when it is pivoted. In order to seat the flat head screws flush in the hinge, the screw holes are recountersunk from the back.

To make a sharp-cornered bend, the mark on the work is positioned in alignment with the center of the hinge. A sharp-cornered block of hardwood is clamped on the line. The flap is raised to form the bend.

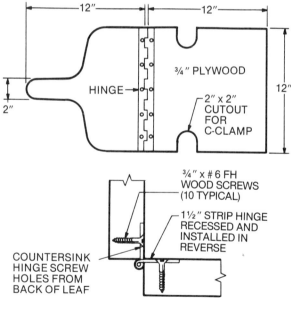

HINGE DETAIL

Bend line on the metal is placed directly over the center of the hinge. The flap is raised to make the bend. When three or four sides need to be bent, the back-up block must be sized to fit within the folded sides.

Here a T bevel is used to check the progress of an angular bend to assure the correct angle.

This is a simple makeshift setup for bending thin aluminum sheet. Bend line and the edge of the hold-down board must be in line with the sharp edge of the table.

PART II

Tips and Techniques

IN THIS SECTION you will find handy shop tips—practical solutions to woodwork problems and procedures for achieving improved results with tools and related operations. The ideas presented here will show you better and safer ways to perform shop tasks—and, in some instances, easier and faster methods.

14. SAWING

Sabre Saw a Dovetail

While the sabre saw is by no means suitable for producing a cabinetry-quality dovetail joint, it can be used to cut this joint with reasonably good results for rough carpentry constructions. The cuts are made freehand with a saw that has a base that can be adjusted for a bevel cut.

Dovetail (upper right) is cut first with the saw base in the normal, right-angle position. Saw is then tilted alternately left and right to make the beveled endwise cuts for the socket on the mating piece.

The saw base is returned to normal to clean out the waste. It is held with the back raised to dress the bottom of the socket clean to the angled corner.

Guided Curved Cut

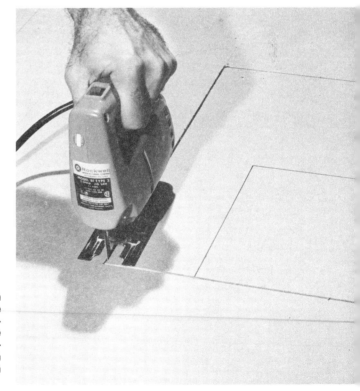

Long sweeping curves can be cut true and smooth by tracking the sabre saw against a curved guide. A flexible stick is tack nailed to the work in the desired shape, offset by an amount equal to the distance from the blade to the edge of the saw base. Note: since the side of the saw base is straight, only the front and back corners will ride against the guide. This will result in a minor offset between the cutting line and the actual cut. If necessary, this factor can be measured and allowance made when positioning the guide.

No-Waste Inside Cuts

When a piece of work is to be cut without a lead-in from an edge and both pieces are to be used, it is possible to make the pierced cut without boring an oversize blade-access hole. Instead, a drill bit of the same diameter as the thickness of the blade is used to bore several closely spaced holes on the cutting line. Insert a loose coping saw blade into one hole, then carefully cut through the others to form a narrow slot which will pass the sabre saw blade.

Mitered Edge Half-Lap

Because it is stronger and neater, this joint is often used in deck construction in lieu of the toenailed butt joint.

This useful joint can easily be cut in large planks with the sabre saw. Saw is tilted to the desired angle; and two parallel cuts, spaced equal to the thickness of the stock, are made through half the width. Saw is guided against a clamped, straight-edged board. An identical cut is made in the mating member—the bevel angle of the saw is not changed.

Chisel is used to clear out the waste.

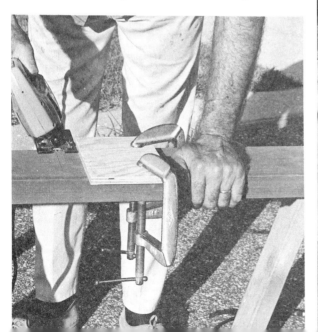

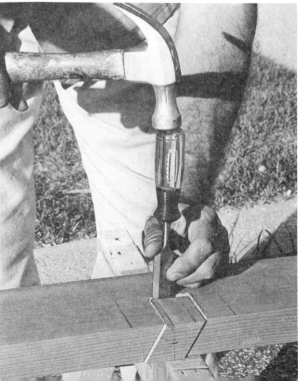

Radial Corners

Whenever feasible, opt to design round inside corners in your projects. They make cutting and sanding much easier. The holes are bored first with a drill or hole saw, then a straight guide is set up so you can saw tangent to the curve.

Truing Edges for Gluing

The edges of boards, particularly long lengths, are seldom straight enough to result in good overall contact for edge-gluing a perfect joint. The problem can be solved easily with a portable saw in the following manner: place the boards on sawhorses with the edges butted together. Clamp them in place together with a straight-edged guide positioned so the blade will cut through the joint line to remove some stock from both edges. This will result in two matched edges. If the blade does not contact both edges on the first pass due to areas with extra-wide gaps, simply butt the pieces together again and make another pass.

The saw is guided to cut through the center of the butted irregular edges of the stock.

Cut Masonry

Bluestone and slate, often used in home improvement projects, can be cut to size with a portable saw equipped with a masonry cutoff wheel. Be sure to wear protective goggles for this operation.

Sawing Ceramic Tile

Sabre saw fitted with a carbide-grit blade can be used to cut ceramic tile. Irregular shapes and cuts to inside corners (which are not possible with regular tile-cutting fixtures) are easily done with this saw and blade combination. The blade does not have conventional teeth; its cutting edge consists of tiny chips of extremely hard tungsten carbide.

Scroll-Saw Pointers

Following the cutting outline with precision on the scroll saw is a skill that is acquired with practice, of course. But you will speed the process if you observe some simple rules: when cutting irregular shapes, feed slowly and steer the work so the side of the blade is continuously tangent to the cutting line. Direct the feed pressure forward into the teeth rather than into the side of the blade. Broken blades and less-than-good work frequently result when an attempt is made to cut sharp corners in a continuous, uninterrupted pass. Normally, there is no need for such action. The correct procedure for handling corners is shown here.

When the main waste has been removed, return to the corner to make the single, clean-out cut to the point.

This is the best way to treat an acute inside corner. Make the first pass directly into the corner. Back out a bit, make an easy turn to the second line, and continue the main cut.

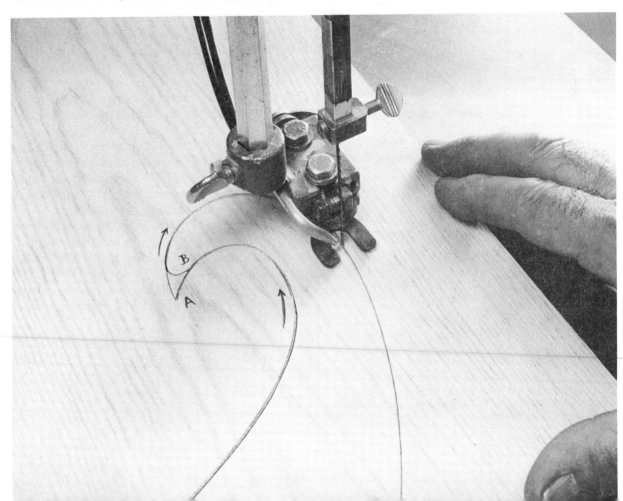

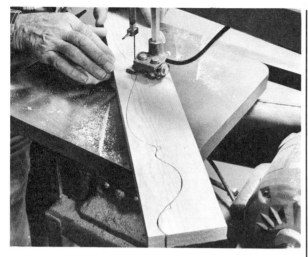

If a curved cutting line includes small, intricate breaks, simply by-pass them and make a continuous run to cut the larger, sweeping curves first.

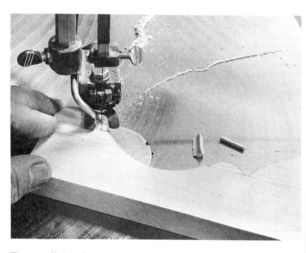

The small details are cut separately, approaching the corners from two directions.

This is a path that can be followed for cutting an outside and inside right-angle corner. The loop into the waste results in a sharp corner in the initial pass.

Angle Sawing

One of the main features of the scroll saw is its ability to permit inside or pierced cuts. These are cuts that start and end within the material without a lead-in cut from the outside edge. When such a cut is made with the table tilted to produce a slight bevel, the space left by the saw kerf will allow the inner piece (or pieces) to jam tight when pushed through to the point where the beveled walls make contact. The procedure is an economical way of building up hollow turning blanks for lathe projects, making raised letters, or to rough form blanks for carved model-boat hulls.

The degree of "telescoping" that results depends on the thickness of the blade and its resultant kerf, the amount of bevel, and the thickness of the stock. In general, a 3-degree bevel is a good starting point for a test. It should be noted that for this kind of cutting it is essential that the work always be kept on the same side of the blade. The piece will be ruined if it is swung from one side to the other, because the bevel angles will change.

Roughing a model-boat hull with this method is not a new idea. The standard practice has been to bore one blade-entry hole for each segment to be cut. This method was improved upon in making the hull shown. Instead of a single hole, two were drilled at the back corners of each section. To make the series of cuts on the right side, the table was tilted 4 degrees to the left, and the cuts were made from the back to the point. The table was then tilted 4 degrees to the right in order to make the cuts on the left side of the hull. The cuts across the back, from hole to hole, were made with the table in the same right-tilt position to sever the sections.

The advantage of this modified technique of two starting holes and opposing tilts was that inasmuch as no sharp turns needed to be made, a relatively coarse 10 TPI blade (.020″ thick by .110″ wide) could be used. This resulted in much faster and easier cutting of the 5/4-inch-thick hardwood stock.

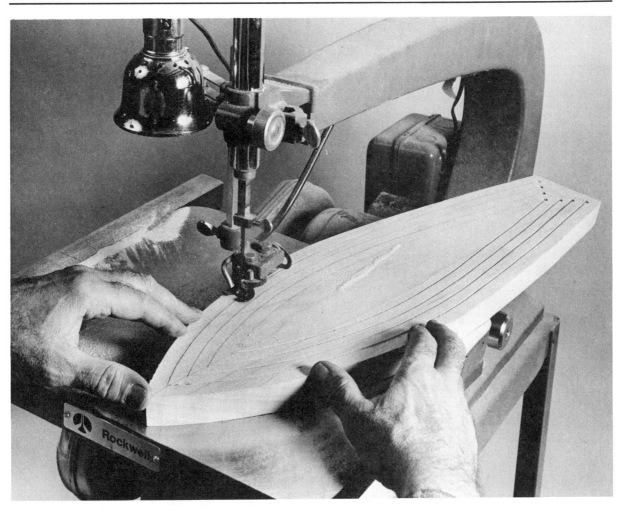

Angle sawing to produce a prehollowed model boat hull. The table is tilted 4 degrees. Cuts are made from each angled hole to the point.

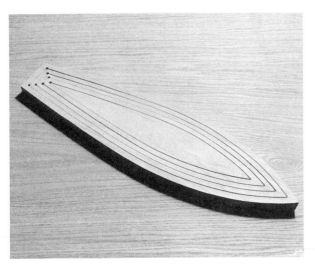

Blank completely bevel-cut.

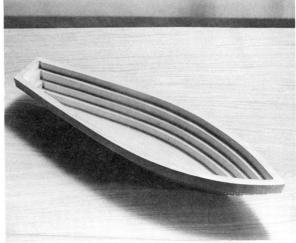

Expanded segments ready for carving. Gluing the self-jamming segments is done without clamping. Glue is applied respectively to the upper and lower areas that make contact. A tight joint is formed when the pieces are pushed through.

Through Mortise and Tenon

The scroll saw is essentially a machine designed for freehand cutting of irregular shapes, but it can also be utilized for controlled precision sawing. The mortise-and-tenon joint is an example. It is made with the aid of the rip guide shown on page 19. Use the following procedure.

Lay out the rectangle for the mortise and bore a through hole at each end equal in diameter to the width of the cutout. A ¼-inch wide 7 TPI (teeth per inch) blade is best suited for this work. The cutout must have precise, straight walls so the cuts

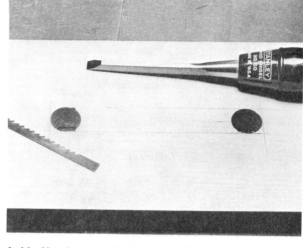

A chisel has been used to shave small flats in one hole (left) to allow the blade to start the cuts tangent to the lines.

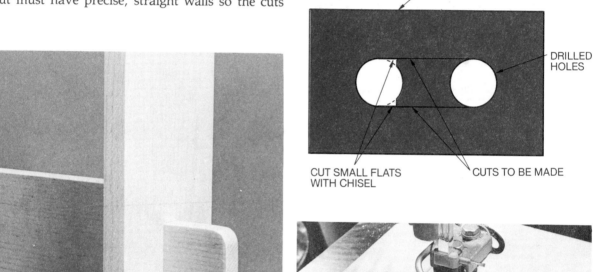

WORKPIECE

DRILLED HOLES

CUT SMALL FLATS WITH CHISEL

CUTS TO BE MADE

This neat joint illustrates the capabilities of the scroll saw when used with controlled cutting methods.

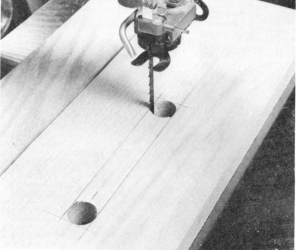

Rip fence is locked in place when the work is positioned with the blade snugged against the flat. The blade guide is up for clarity in the photograph.

Blade is rotated so the teeth face the opposite direction after the long cut has been made. The blade-guide assembly is removed for this phase.

Work is pulled backward to make the small cut to the end beyond the hole.

Work is taped to the fence for the scrape-sawing operation to square the ends. The fence and work are moved back and forth until the surface is smooth and true.

must start tangent to the holes. The relatively wide blade cannot hug the wall of the round hole closely enough to accomplish this, so a chisel is used to cut two small flats in one hole, tangent to the cutting lines.

With the rip fence set up to guide the work with the blade snug against the flat, make the first straight cut through to the bottom cross line. When this has been done *do not* move the fence. Instead, remove the blade guide. Remove the blade and rotate it 180 degrees so the teeth point to the rear. Guiding the work against the fence, pull it backward to complete the cut past the hole and to the upper cross line of the rectangle. Shift the fence and repeat the procedure to make the second straight cut. Begin in the reverse order, cutting into the top corner first; then turn the blade around to its normal position and reattach the blade-guide assembly. Make the cut and remove the cutoff waste. The result will be a slot with radial tips remaining at each end.

The purpose of using the reverse-blade technique is to avoid the need to shift the guide fence as would be required in order to make the corner cuts in the normal manner because the shift could result in a misaligned cut.

The ends of the opening are squared to perfection by a slide-sawing action. The rip-fence locking screw is loosened as the fence must be free to slide for this operation. The work is positioned so the teeth of the blade touch the end cross line. The work is taped to the fence. The power is turned on, and the fence is moved back and forth allowing the blade to scrape the end clean in a sideways motion. In the final strokes the fence should bear firmly against the supporting angle-iron edge. Both the rip and crosscut members of the jig are used to cut the tenon and complete the joint.

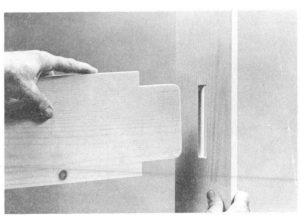

Completed joint, ready for assembly.

Ganged Blade Edge Half-Lap

The edge half-lap joint slots can be made on the scroll saw by making two guided parallel cuts spaced equal to the thickness of the stock. This can be done in stock of any thickness within the capacity of the saw. However, in stock up to 1/4 inch thick you can cut the slots in a single pass by ganging several blades to obtain a wide kerf that will be equal to the thickness of the stock.

The procedure is feasible on the Rockwell scroll saw, which has a lower chuck capacity of 1/4 inch. The upper chuck does not have this capacity so it is not used for this operation. Other scroll saws with a similar, lower chuck capacity can also be used for this application. The required number of blades are locked in the chuck; then they are wrapped around the top with tape to prevent whipping.

If the workpiece size permits, a strip is clamped to the table to guide the work for a true crosscut. If the workpiece overhangs the table, the guide strip is attached to its underside and bears against the table edge to guide the work.

Ganged blade trick permits cutting wide slots in a single pass for edge half-lap joints in 1/4-inch stock.

Blade pack should consist of similar blades so that the set teeth will nest together neatly. A straight-edged guide is clamped to the table for this cut.

Work overhangs the table here, so the guide is clamped to the work and rides against the table edge. If the weight of the clamps causes the work to belly upward at the blade location, the hold-down foot should be attached.

Coped Molding Joint

The coped molding joint is a good one to know about for use in certain situations where the mitered joint is not feasible. In this joint the second member is cut with a profile that will match and butt up against the first. The cut is easily made on the scroll saw.

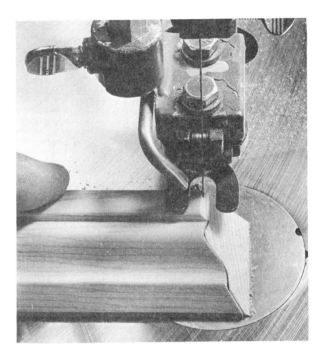

Regular, straight 45-degree miter is cut on the end of the molding to reveal the exact contour outline on the face. A straight 90-degree cut is made following the outline.

New edge is a duplicate of the face at a right angle.

When carefully cut, the coped piece will butt perfectly.

Sawing Ceramic Tile

As with the sabre saw, the carbide-grit blade can be utilized in the scroll saw for problem materials such as this glazed, ceramic floor tile. A hardboard pad taped down will prevent the table from being scratched by the abrasive tile dust.

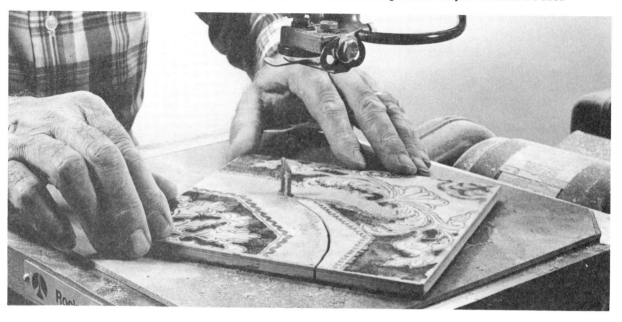

Inlay Strings

If you have ever tried to cut very narrow, irregularly shaped inlay strings in the conventional manner you will appreciate this technique. Simply use two identical blades separated the desired amount by wood-strip spacers taped in between, above and below the cutting area. The single cut will form twin kerfs resulting in a parallel-edged strip.

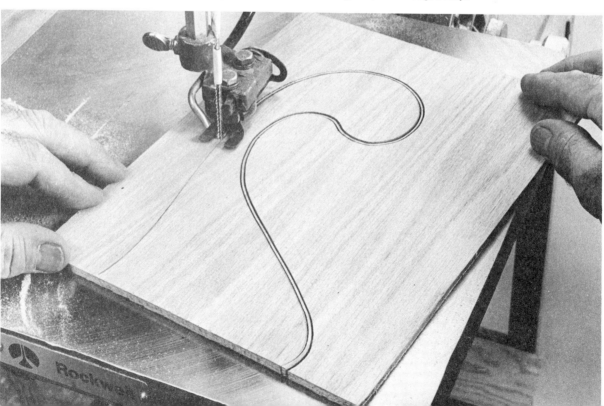

Band Saw Safety Tip

Do not press your luck when sawing small pieces on the band saw. Here is a safe and sound way to handle an otherwise dangerous cut.

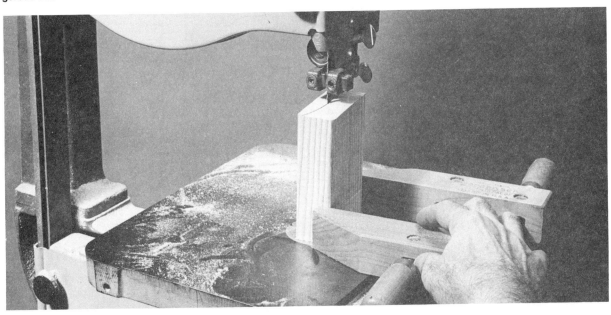

Compound Cuts

Compound cuts shaped on three or four sides are easy to make with the scroll saw. The trick is to make careful cuts so the waste cutoffs remain intact. They are taped back into place and outlined for the subsequent cuts.

Pre-Form the Radii

The technique of boring holes to pre-form the radii in sawing operations is especially advantageous on band saw work—particularly if tight curves are involved. The curves will be perfectly formed and will require only minimal sanding.

The importance of accurate placement of the holes is obvious. When the outline is composed of straight lines, centermarking of the holes is a matter of making simple measurements. With irregular curved outlines the use of a circle template will prove very helpful in selecting the correct hole size and in centermarking for drilling.

Workpieces with drilled holes of various diameters, and ready to be cut out.

Sawing is a lot easier when sharp curves can be avoided.

A circle template is used to determine the drill size that will best fit the curvature. Center cross lines on the circles ensure accurate centermarking for drilling.

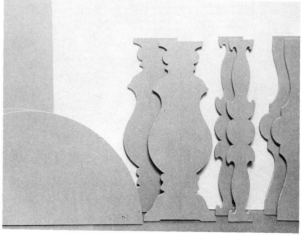

Sawed workpieces will need least sanding in the areas of the drilled holes.

Multiple Small Cutoffs

Make small cutoffs of uniform length with this simple setup: clamp a wood-strip fence to the table, parallel to the side of the blade and spaced for the length of the desired cutoff. Butt the work and a wide guide block against the fence. The blade guard is raised here for clarity in the photograph. It should be lowered for safety.

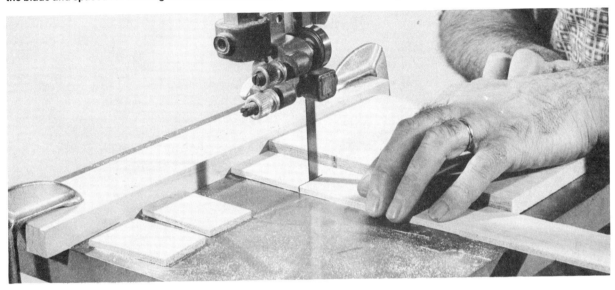

Sawing Plastic Sheet

Cutting lines on deeply textured decorative plastic panels are difficult to mark properly for sawing. The solution for obtaining clear, easy-to-follow lines is to draw the pattern on paper and tape the paper to the plastic.

Sawdust Relief Slot

Sawdust will always collect against the fence of a radial-arm saw after a cut has been made. Unless every bit of this dust is removed the next piece of stock to be cut will not butt properly against the fence; this will result in an untrue cross cut.

You can avoid the bother of constant brushing by attaching little stand-off spacers to the lower front edge of the fence. The gaps between the fence and table will allow the sawdust to fall through, thus preventing any accumulation.

Nail 1/4-by-3/4-by-1-inch spacers about 6 or 8 inches apart. Make certain that a block is not located in the path of the blade.

Notice how the freshly made cut resulted in no accumulation of sawdust against the fence.

Splined Miter Groove

When cutting the groove for a splined miter joint, the best working position is with the beveled edge of the work against the fence. But the groove may go slightly off course if the sharp pointed edge of the bevel should break as it rubs against the fence. To avoid the problem, shave the point back about 1/16 inch beforehand so a sound edge rides against the fence. This should have no adverse effect on the finished joint because, in all probability, the sharp corner in the assembled work will be eased anyway.

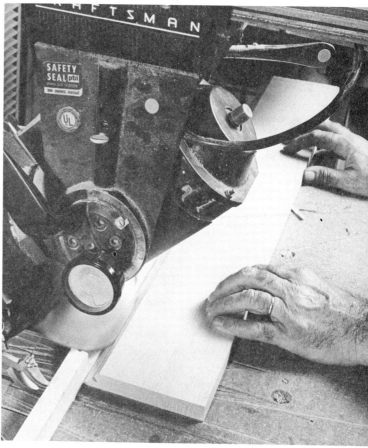

A flat of about 1/16 inch width will suffice to allow the work to ride firmly against the fence. Notice how a recess has been cut in the fence to provide clearance for the blade.

Avoid Erratic Half-Laps

The half-lap joint is a relatively easy one to cut as it merely requires reducing the thickness of the stock an equal amount in the mating pieces. When the results are not up to par the fault may lie not with an improper blade height adjustment but, instead, with warped lumber. A slight belly in the face of the work resting on the saw table will cause it to rise at

the blade location. Consequently, the resulting cut will be too deep.

Avoid the problem by forming a habit of always observing how the stock is resting before making the cut. If a gap exists, the work should be clamped to obtain good contact with the table. If clamping is not feasible, prop up the far end.

Gap between this workpiece and the table will result in too deep a cut.

Far end of this workpiece is propped up, resulting in proper table contact and a cut of accurate depth.

When working with stock up to ³/₄ inch thick (which has some flex), another procedure may be employed: two pieces of right-

angle outside corner molding nailed into the fence will serve as hold-downs to ensure close contact of the work to the table.

Sizing Notch Cuts

Workpieces requiring a number of similar-sized notches or dado cuts can be cut to exactly matching widths by employing a simple trick. Install a new, kerf-free section of back-up fence that stands about 1/4 inch higher than the work. Begin by carefully measuring and cutting the two outside bordering cuts in the work to establish the width of the notch. When the second kerf cut has been made, clamp the work in place on the saw table. Then loosen the fence-holding clamps and shift the fence to align its original kerf cut to the first kerf cut that was made in the work. Lock the fence in place. Remove the work and make a second kerf cut in the fence. Thereafter, when one border cut has been made, shift the work until the kerf lines up with the extra kerf in the fence. Visual alignment of kerf to kerf will result in identically spaced border cuts. After all these cuts have been made, the pieces are returned to the saw to clear out the waste.

Accurately spaced kerf cuts in the first workpiece are used to space the second kerf cut in the fence. Subsequent border cuts in the work are made by visually aligning the kerf cuts, which will be relatively easy and quick.

Pushing the Work

Use two push sticks when trimming narrow strips. Position the second one on the side of the strip just before the blade to serve as a hold-in. As the end of the cut approaches, slide it back to engage the rear of the strip as shown.

This is how to cut a deep bevel in small work the safe and easy way: partially drive two nails to serve as handles. Stand off to the side of the saw and hold the work with two fingers of each hand straddling the fence, as shown.

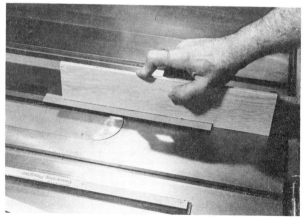

Push sticks need not be of the conventional handled shape. This one serves also as a hold-down in making a rabbet cut in this delicate workpiece.

Adjustable Push Stick

This trigger-grip push stick is somewhat sophisticated. It wraps over the fence to ride steadily and has a built-in hold-down that adjusts for stock of varying thickness up to 1 inch. When constructing it be sure to bore the hole for the dowel *before* cutting the contour. Coat the snug-fitting dowel with wax for easier sliding.

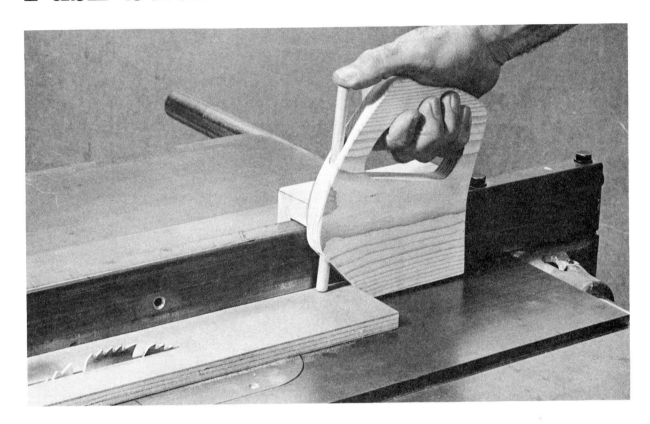

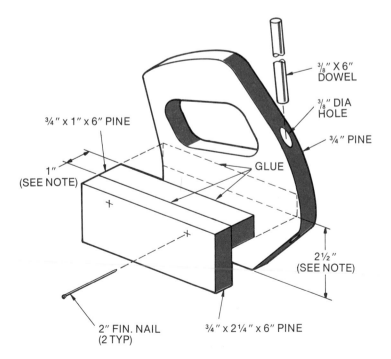

¾" x 1" x 6" PINE

1" (SEE NOTE)

2" FIN. NAIL (2 TYP)

¾" x 2¼" x 6" PINE

GLUE

⅜" X 6" DOWEL

⅜" DIA HOLE

¾" PINE

2½" (SEE NOTE)

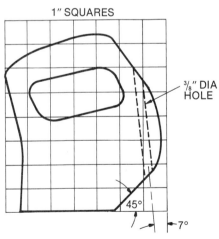

1" SQUARES

⅜" DIA HOLE

45°

7°

<u>NOTE</u>
DIMENSIONS SHOWN ARE FOR A 1" WIDE x 2½" HIGH RIP FENCE; MODIFY DIMENSIONS AS REQUIRED

Cutting Bench Dog Holes

Construction plans for woodwork bench tops usually specify a row of round holes to house ordinary bench dogs (which are simply round wood pegs with square or round shoulders). Professional cabinetmaker's bench tops utilize far superior spring-tensioned, square-shaped steel bench dogs that fit into square holes. This type can be set to project at varying heights, holds firmly, and will not twist in the hole.

When you become involved in constructing a workbench, you might consider using the heavy-duty dog. Of course, the prospect of making a dozen or more square holes with side shoulders through three or four inches of tough hardwood could be discouraging—but there is an easy way.

Since bench tops are made by laminating stock side by side, the holes can be pre-formed in one of the slab members prior to assembly. This is done with the radial-arm saw, as shown.

Blade height is adjusted to cut a kerf equal in depth to the thickness of the dog. Two through kerf cuts are made to set the width of the dado. Partial cuts are also made to partly form the shoulder.

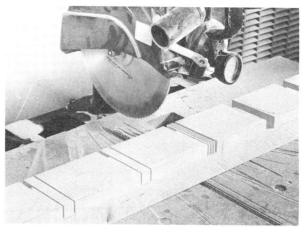

Repeated kerf cuts are made to form the dado. A dado head could also be used.

Bench dogs are an essential feature of a good cabinetmaker's workbench. This bench is homemade.

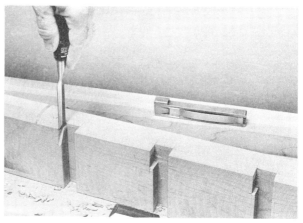

Chisel is used to complete the shoulder, which serves as a stop for the dog. Notice the spring in the top-quality bench dog.

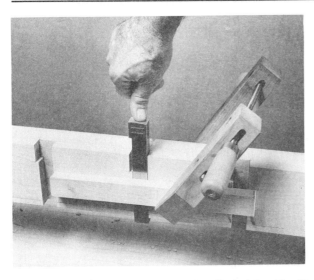

Each slot is tested in this manner before final gluing. The fit should be slightly tensioned.

Dadoes become square holes when the adjoining slab member is added.

Spacing Kerf Cuts for Bending

Bending wood by kerfing will result in graceful, symmetrical curves only if the kerfs are equally spaced. An easy way to accomplish this is to cut a second kerf in the fence spaced equal to the spacing required in the work. As each kerf is cut in the work it is shifted to line up with the guide kerf in the fence.

Equally spaced kerf cuts will allow the flexible piece to bend gracefully. The thickness of the resulting shell governs the degree of curvature possible.

Gang Cut to Save Energy

Save time and electrical energy by gang cutting when similar cuts must be made in a number of pieces. Reduction in time spent in setup and alignment of individual pieces translates into reduced running time of the motor and, of course, uniformity of the cuts. When possible, use clamps to hold the bundle together.

Gauging Small Crosscuts

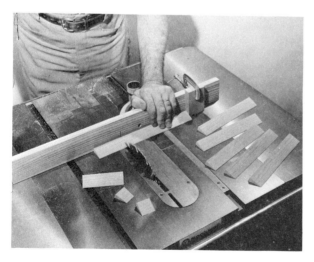

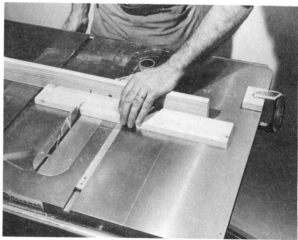

Gauge multiple crosscuts simply by clamping a stop block to the miter fence.

If the work extends beyond the fence, the stop can be clamped to the saw table. Be sure to locate the stop so the work is free of it before it engages the blade.

Increased Bevel Angle

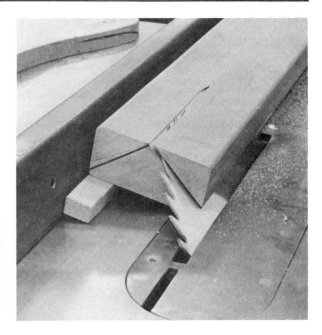

Here is a good trick to keep in mind when you need to make a bevel cut at an angle greater than the tilt capacity of the saw. The steep angle in the example shown would be in the range of the saw's tilt capacity if the piece were to be run through on edge; but this would be an unsafe cut because as the cut progressed the supporting base of the work would be gradually diminished. The problem is easily solved by tacking a strip of wood to the underside to increase the relative tilt, as shown. In this way the workpiece has firm footing throughout the cut.

Ripping Round Stock

To make a safe and accurate lengthwise cut on round stock do this: butt the round against a piece of flat stock on a flat surface, then tack the two together with a few beads of hot-melt glue along the top edge where they join. When the cut has been completed snap the pieces apart. Glue beads will usually pop off clean. If not, a chisel will do it. Attempting this cut without such an aid could be quite dangerous.

Strip Saver

Small thin slices of wood ripped on the table saw usually get swallowed into the saw if the space around the table insert is larger than the strip. Overcome the problem by taping a piece of thin plywood over the depressed blade, then turn on the power and elevate the blade until it cuts through. With no space around the blade the small pieces cannot fall through.

Support Thin Stock

When cutting thin stock with the rip fence positioned beyond
the table surface, a strip of wood clamped under the fence will
prevent the work from slipping under and fouling the cut.

Auxiliary Fence

Attach an oversize high fence board to the table-saw rip fence
to obtain better control in feeding wide work over the blade on
edge. Use a feather board (page 34) to apply side pressure.

Notching Corners

Clean-cut corner notches can be made with two right-angle table-saw cuts with the blade set high in order to minimize the length of the radial kerf on the underside. In this operation it is important to cut in a sequence that will let the cutoff be on the free side of the blade rather than between the blade and the fence. Otherwise the loosened scrap, trapped between the blade and fence, could jam and kick back.

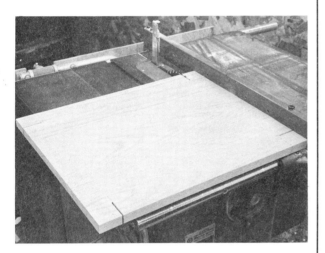

First cuts are made with the notch areas between the blade and fence. This panel will have four notched corners.

Second cut for the notch should place the cutoff on the free side of the blade.

Which Cut First ?

When multiple bevel cuts are to be made in a workpiece, it is important to consider carefully the order of cutting that will result in a safe operation.

Making these cuts first was a wise move in this situation because the base is fully supported.

As final cuts are made the base is still fully supported. If these cuts were made first the small surface would be down during the final cutting, and the piece would be likely to tip.

Panel Raising Basics

Bevel-raised panels are widely used on cabinet and cupboard doors, shutters, and interior and exterior doors to add an interesting design feature. The panel is set into grooves of a frame consisting of two stiles and two rails. If two panels are used, one above the other or side by side, a third center rail or stile is used as a divider. A variety of corner joints may be used to assemble the frame. The stub mortise-and-tenon joint shown is widely used for light constructions. When greater strength is desired, two dowels may be added to reinforce the joint.

Plywood is generally avoided for the raised panel because the bevel cut will expose the plies; these show up when a finish is applied and are unat-tractive. Industry uses plywood extensively for the panel, but this plywood is specially made for the purpose and is not available on the consumer level. It is constructed with inner plies sized so the bevel cut can be made within the single ply. Kiln-dried solid wood is suitable for the panel provided it is warp-free.

A smooth-cutting hollow ground miter or planer blade should be used as this will produce a surface requiring little sanding. The bevel angle is usually 15 degrees, but this can be varied to suit the dimension of the stock. The overall dimensions of the panel should be about $1/16$ inch less than the spaces between the bottoms of the frame grooves.

Basic parts of a raised panel. The center rail for this two-panel unit is twice the width of the top and bottom rails.

End bevels are cut first using a simple block jig for stability. Fence is positioned to leave a ³/₁₆-inch-thick outside edge. This can be adjusted to your plan.

When grooving the short rails with a dado head, the large space around the blade insert can present a problem unless you apply downward pressure on both the front and back of the piece as it passes over the blade.

Side bevels can be fed freehand, but for positive control it is better and safer to stand on side of saw facing the fence. Make sure fingers are clear of blade.

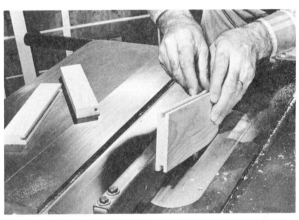

This is what can (and probably will) happen if downward pressure is applied only at the rear of the piece. Damage to the work and possibly serious injury to the fingers can result.

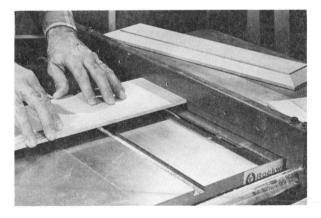

The set-back bevel cut will leave an angular shoulder which is trimmed square with a shallow cut. A test piece (at right) is essential for making trial cuts to ensure proper fence and blade adjustments for this critical cut. Shoulder is optional—the bevel can be made flush to the surface.

A block jig similar to the one used for beveling the ends of the panel is a big help in cutting the tenons. The Tenoning Jig shown on page 35 would serve nicely for this.

Built - In Joints

It pays to be on the lookout for the opportunity to pre-form joints during the initial stages of construction or prior to the joining of sub-assemblies. Quite often much time and effort can be saved with advance planning, and finer work may result as well. Sometimes the shortcut can be worked in by making a minor design modification or it may be chanced upon during construction. Two such situations are illustrated here as examples of the possibilities. While the specifics will be of limited value to you, they may serve to get you to think along the same lines in your project activities.

The original plan for the table shown here during construction specified legs to be made from individual pieces of solid square stock. This would have required difficult and painstaking hand-chiseling to form the inside corner notches to house the rails. This detail was modified, and the legs were made by gluing up stock in two stages to result in preformed, clean-cut inside corner notches. The photo sequence shows how this was done. The other situation involved a table slab that required two large blind mortises. Here the joints were cut prior to final glue-assembly of the slab.

The plan for this table project originally indicated corner notches to be cut in solid stock for the legs. This would have been a trying task with questionable results.

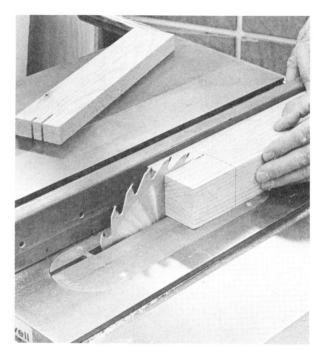

Two lengths of stock, equal in combined thickness to the width of the notch, are glued together. An end cut is then made. This cut produces a kerf undercut on the bottom face. The cuts are therefore alternated left and right on the other legs so the kerfs will be concealed on the inside when the third piece is glued on. This precaution results in symmetrical-face/edge-grain orientation—an important consideration if the end-product is to be clear-finished.

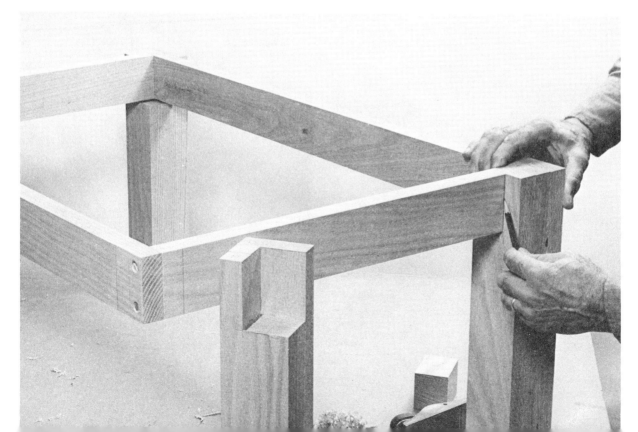

A crosscut is made to clear out the waste, thus forming an open notch.

Dadoes are cut into a partially assembled slab.

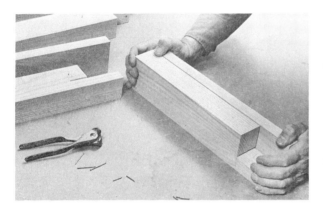

Headless nails serve as pins to align the pieces for gluing and to prevent them from sliding when clamped. Nails are set into pre-drilled holes with the points projecting about 3/16 inch. Parts are pressed together to obtain nail-point registration holes. The third piece forms the inside corner.

Dadoes become blind mortises when the remaining pieces are glued to the slab.

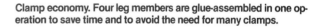

Clamp economy. Four leg members are glue-assembled in one operation to save time and to avoid the need for many clamps.

Aligning Dadoes

Aligning the work to make a series of precisely aligned dado cuts will be easy if you install a fresh fence board to the miter gauge. Elevate the blade to make a cut about ¼ inch higher than the thickness of the stock to be worked. Make the cut into the backboard, then adjust the blade to the height required for the dado. Mark the dado locations on the top edge of the work and simply align these marks against the clearly visible notch in the fence.

When the rip fence is used to cut dadoes, make a trial cut in scrap stock; then back off the test piece to the front of the table insert. Mark the insert to indicate the cutting path. Align the cutting marks on the work with the marks on the table, then move the fence to the edge of the work and lock it in position. This procedure is particularly useful when using a wobble-type adjustable dado head, because with this type of blade there is no point of reference from which to measure.

Crosscut Lineup Guide

Lineup of the cutting mark on your work to the blade will be easy and accurate if you mark the cutting line of the blade on the saw table. Apply a strip of thin tape to the table, then make a cut in a scrap of wood. Without shifting the scrap, slide the miter gauge back to position the cut end onto the tape. Mark a fine line against the edge of the wood. This will indicate the inside face of the blade. When a blade of a different thickness is installed in the saw, simply mark with a different marker.

Trim a Drawer Bulge

A drawer with a bulging side can be trimmed level with a dado head on the table saw. Tack nail two strips of wood to the side so the drawer stands off the table surface about ½ inch. Adjust the blade height to the level of the part of the side that is correct. Make repeated passes to shave off the high spot. A strip of tape applied to the table in front of and behind the blade will indicate its location and will help to avoid running the guide sticks into it.

Chain Saw Joinery

While it will not produce joints to strike the fancy of the fine woodworker, the chain saw can be used to make quick work of cutting somewhat crude but serviceable half-lap and open mortise-and-tenon joints in heavy lumber for rough carpentry constructions (outdoor structures, rustic fences, and the like).

End laps and tenons are readily formed by making cuts at right angles. To notch out middle laps and through mortises, a series of kerf cuts are made in the same manner as is done with the table saw in shop work.

A good part of the joinery in this structure was done with the chain saw.

Series of kerf cuts are made halfway through the stock to form a half-lap notch. A horizontal cleat (out of view at the back) is used to guide the depth of cut.

Cutting a through mortise in a 6 × 6 post. The cleat across the back serves as a depth-of-cut guide. A chisel is not the right tool to use for removing end-grain slices, so they are nibbled away by feeding the saw at a slight angle after the vertical cuts have been made.

Chisel is used to snap off the waste slices.

Cutting an angled end lap in heavy stock is easy with this tool.

Matched Compound Miters

The operation of cutting identical compound miters on two ends of a piece of stock may stump you when you use the table saw: the blade can tilt in one direction only. But there is a technique—simply flip the work bottom side up and rotate the miter gauge 180 degrees for the second cut, as shown.

Miter gauge (above) is set to the required angle, and the blade is tilted to cut the compound miter. Gauge is in the normal position for the first cut.

Miter gauge (below) is placed in the second slot and turned front to back. The workpiece is flipped over to make the second, matching cut.

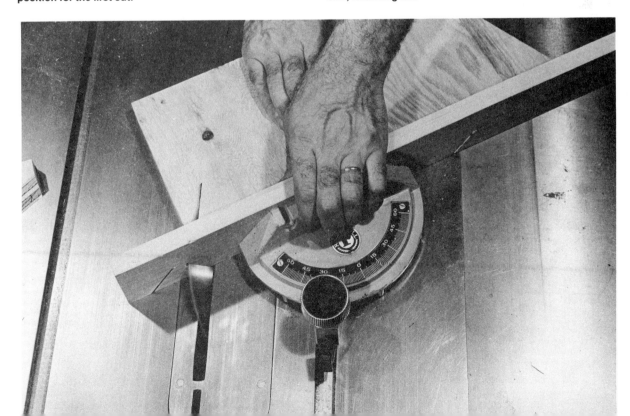

Ripping Straight

Although boards with surfaced edges presumably have perfectly straight edges, actually they rarely do. When critical work demands perfectly straight edges in long boards that are to be ripped to width, a discrepancy in the edge riding the rip fence will transfer to the new cut edge. The best way to obtain a straight cut is by tacking on a true straightedge, slightly overhanging, to bear against the fence.

In checking over the materials on hand for something to use for the guide, you may find it difficult to come up with a likely candidate. Solve the problem by obtaining a length of aluminum extrusion—a wide sliding-door track or an H section such as that used as the divider for heavy wall panels. Both are usually quite straight. Drill a series of holes to receive finishing nails and keep the piece on hand for those special jobs.

An H-section aluminum extrusion makes an excellent straight-edged guide. Holes for the nails are pre-drilled.

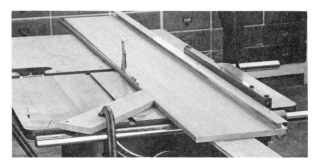

The overhanging edge of the guide rides against the saw fence to true the working edge of a board. Subsequent rip cuts from this edge will also be true. The feather board (page **34**) hold-in is essential for good results with bulky boards.

Long boards should be supported fore and aft during ripping. This makeshift set-up serves well and allows room for standing close to the saw.

15. SURFACING AND SHAPING

False Drawers

Large drawers are sometimes required in a project, but they tend to be monotonous in appearance. The appearance can be improved by cutting narrow grooves to give the illusion of several drawers. This is usually done by cutting shallow saw kerfs in the surface. When it is desired to butt right-angled grooves, as shown here, the saw is used for the initial through cut. The stopped grooves are then cut with a router.

The single, large drawer under this countertop is made to simulate three drawers with vertical grooves that are cut into the front panel.

Router with a veining bit is guided against a clamped block to make the stopped cut up to the saw kerfed groove.

Rounding a Dowel End

Try this trick when you want to round a dowel end to perfection. Insert a hanger bolt centered in the end of the dowel and chuck it into a portable drill. Spin the workpiece against the rotating sanding disc while constantly swinging the drill in an arc. To form a taper, simply hold the drill in one position at the desired angle. The procedure can also be used for rounding the ends of square stock.

When fast stock removal is desired, position the work so the rotations oppose each other. Position the work so the rotations are in the same direction for final smoothing.

Sizing Dowels

Dowel diameters frequently do not match respective hole diameters precisely. When the fit is too tight you can trim the dowel to size on the drill press in the following way.

Bore a hole of the appropriate size in a scrap of wood, then clamp the wood centered on the table. Grip the dowel in the chuck, with the bottom about $1/2$ inch or so above the hole in the wood. Turn on the power and hold a sanding block at an angle against the dowel end to form a chamfer at the bottom to allow the dowel to enter the hole. Rest the sanding block firmly on the surface of the table and lightly touch the block to the spinning dowel while depressing the quill handle. The dowel will slide through the hole as its diameter is reduced.

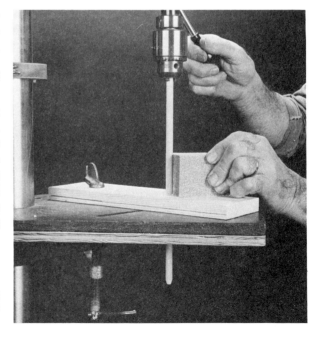

If the dowel is longer than the quill travel capacity, the table is repositioned higher after the lower portion of the dowel has been sanded.

Small Turnings on Drill Press

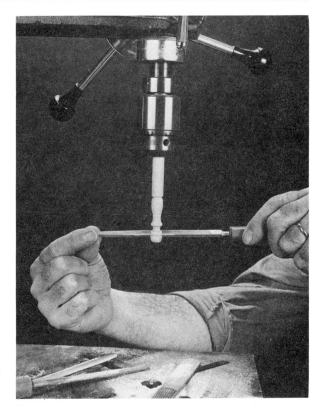

Chuck small lengths of dowel into the drill press and use a variety of files freehand to make ornamental turnings. Operate the drill at high speed for smoothest results and be sure to use a file card often to clear the packed wood from the file.

Edge - Shaping Over Indents

When a self-piloted router bit is used for edge-shaping it is essential that the edge of the work be perfectly flat, without voids or indents, because the pilot end will follow any irregularity in the surface and transfer it to the shaped edge. In work such as the piece shown, where the edges have built-in depressions, the edges can be shaped with a piloted bit by temporarily applying a flush surface.

Strips of ⅛-inch untempered hardboard are applied to the edges to provide a flat surface for the pilot end. The strips are simply snugged into place as shown.

Router bit is adjusted for a greater depth of cut to allow for the added dimension.

Orienting the Grain

When gluing up stock which will require subsequent truing and smoothing with a sharp-edged tool—a plane, a jointer, a spokeshave, and the like—it is extremely important to orient the grain. If this precaution is not taken a frustrating problem will arise: unsightly surface gouges will appear in those areas that are tooled against the grain.

You can sometimes determine the "downhill" direction of the grain by visual inspection. But when you are not sure simply make a small test shaving with a block plane to determine the smoothest cutting direction. Indicate this with a marked arrow on each piece and glue-assemble them accordingly.

This step takes only minutes, but it can mean the difference between a good job and a poor one. Do not neglect it.

Edge - Planing Guide

If you have difficulty in obtaining square edges with the plane you can solve the problem by attaching a fence. Bore two holes in the base of the plane, close to an edge. Attach a square-edged strip of wood to the bottom by driving two round head screws from the upper side of the base.

This fence will hold the plane true if you bear it firmly against the side of the work as you make each pass. To take advantage of the full tool's blade width, the fence may be attached to the side of the plane.

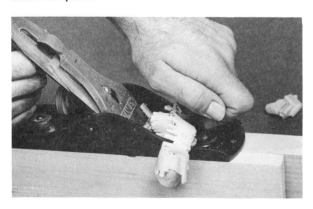

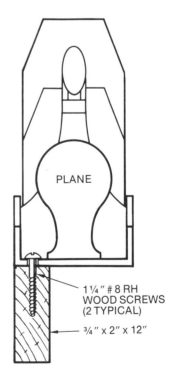

PLANE

1¼" # 8 RH
WOOD SCREWS
(2 TYPICAL)

¾" x 2" x 12"

New Type Rasp

The saw-rasp is a relatively new tool that cuts rather aggressively but leaves a fairly smooth surface. It consists of an array of zigzag-shaped saw blades; the blades are fine-toothed on one side and coarse on the other.

The broad surface of the tool lends to accurate shaping of flats or rounds.

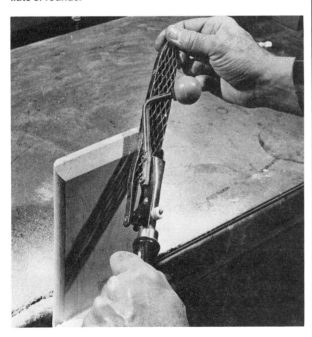

Cornering Tool

This Stanley cornering tool is handy for quickly easing the corners of stock. It is pulled along the edge to round the corner in one stroke. Each end cuts a different-sized radius. It is available in two combinations—1/16-1/8 inch and 1/4-3/8 inch.

Grooved Panel Tip

When you plan to cut accent grooves over the joint lines of edge-glued boards, do not use dowels to reinforce the joints. This will avoid the possibility of cutting through and exposing the dowels. A lap joint or a spline-reinforced butt joint is the better choice. If a spline is used the grooves should be cut off center, closer to the rear surface.

Shaping on a Drill Press

The drill press is sometimes used for shaping operations by chucking in regular router bits. However, since ordinary two-flute router bits are designed to cut efficiently at very high rotating speed, their use in the relatively slow drill press does not produce optimal results.

A special type of bit with nine flutes (cutting edges) effectively offsets this shortcoming and produces rather good results when used for shaping on the drill press or with a portable drill. The greater number of cutting edges in contact with the work per revolution has an effect comparable to a regular router bit rotating at very high speed.

A variety of shapes are possible with this set of nine cutters suitable for use with a drill.

Samples of the cuts made with the drill-driven bits. *Never* use the bits with a router.

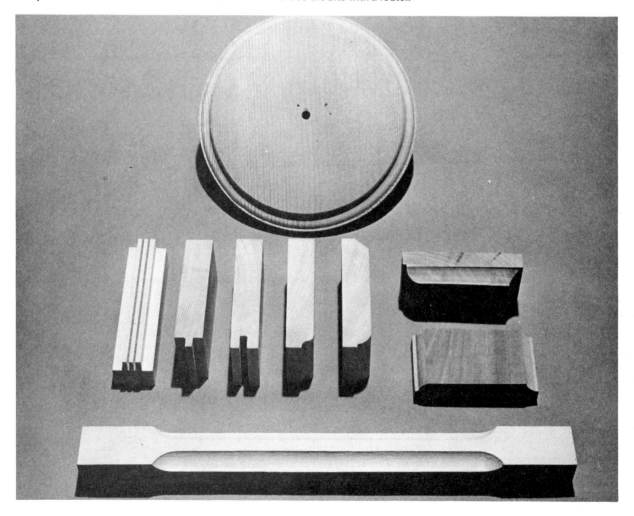

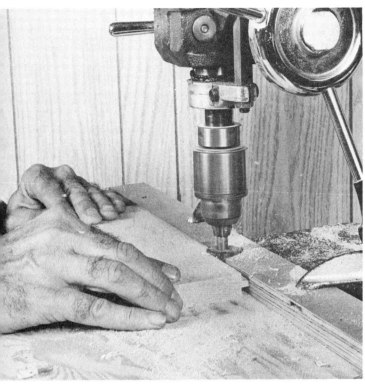

Portable drill held in a drill guide such as this one and attached to the underside of a board becomes a shaper.

Bits are not self-piloted so an edge guide is always required. Clamped wood strip with a cutout around the cutter guides the work for this tongue-forming cut. A mating cutter is used to cut a matching groove for a tongue-and-groove joint.

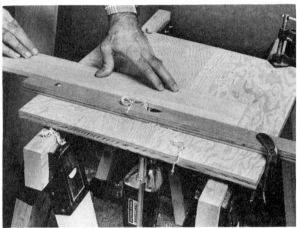

This is the arrangement for shaping with the portable drill. Variable speed drills should be operated at maximum RPM.

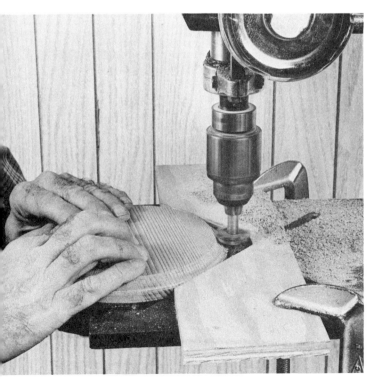

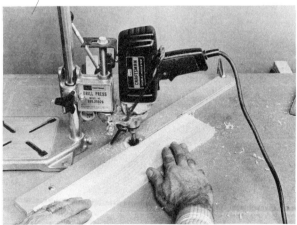

A board with a V notch is used as a guide to shape the edge of a circular disc. Rotating at 5000 RPM, this type of cutter makes 45,000 cutting-edge passes per minute.

Portable drill attached to an accessory drill-press stand will also serve for shaping.

Surfacing Plain and Fancy

The rotary planer is an invaluable accessory for the radial-arm saw. It is primarily a surface planer useful for smoothing or truing broad surfaces and for reducing stock thickness, but it can serve other functions as well.

The unit consists of a die-cast aluminum disc that holds three carbide-tipped steel blades that cut a path 2 inches wide. The arbor is set in the perpendicular position, and the work is pushed along the fence for the planing operation. The planer is moved outward after each cut. Lengthwise rabbets can be cut in the edges of boards in the same manner. For end-grain crosscut rabbets, the work is held stationary and the carriage is moved outward as in sawing.

Because it is so smooth-cutting, the tool is especially useful for making the bevel cuts for panel raising. A variety of decorative surface patterns may be obtained by making repetitive crosscut passes with the cutter tilted.

Another way to make interesting repetitive decorative cuts in wood is with the use of a molding cutter head on either the radial-arm or table saw.

Cutting end of the rotary planer. It is held firmly on the arbor with a nut housed in a hexagonal recess. The carbide-tipped blades can be removed for sharpening.

Planer is tilted to make the beveled cuts for a raised panel. The tool is quite safe because it does not have a tendency to kick back at the operator.

This ornamental strip is produced by making repeated crosscuts with the planer alternately tilted left and right.

Workpiece is clamped in position on the rip fence over the retracted molding cutter head. Power is turned on, and the cutter is cranked up a predetermined number of turns to produce a radial molded cut. Work is shifted and clamped at a new measured position, and the cut is repeated.

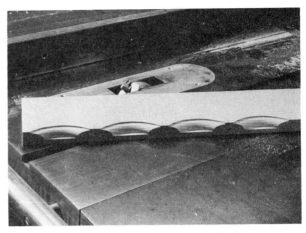

For added variety cutters can be alternated between cuts or second passes can be made with interchanged cutters. A wide variety of designs is possible depending on the spacing of the cuts and the shape of the cutters.

Pin Router Basics

The pin-routing accessory serves to convert the router into a stationary overhead shaper with a simple, but all-important, special added feature—the guide pin. Centered under the router bit and projecting slightly above the table surface, this pin provides complete control of the movement of the work and the resulting cut.

The basic operations that can be performed utilizing the pin are partial edge-shaping and template-guided routing and shaping. When the pin is removed the unit can be used with a fence or for freehand routing.

The pin controls the lateral movement and consequently, the lateral depth of cut. For partial edge-

Samples of the type of work that can easily be accomplished with the pin router accessory.

The pin router unit has a carriage mount that accepts most routers. Three-prong quill feed raises or lowers the carriage on a rack-and-pinion slide. Guide and alignment pins are shown on the work surface at right.

Adjustable pin block is shifted until the guide pin and the chucked alignment pin are centered. See-through guard has been removed for clarity in this series of photographs.

shaping, the smooth, finished straight or irregular edge of the work is moved against the pin. This method of operation permits the use of bits that do not have pilot ends, thus broadening the variety of cuts possible with a given collection of bits. When bits with standard $^3/_{16}$-inch-diameter solid pilot ends are used in conjunction with the pin that has a $^1/_4$-inch diameter, a gap results between the work edge and the pilot. This is an important advantage because it prevents the common problem of edge burn marks caused by the friction heat generated in the fast-spinning pilot as it rubs against the work.

Template-guided routing and shaping operations are the main attractions of the pin-routing capability. Cut to the desired size and shape, the template permits making repetitive cuts on any number of workpieces with ease, speed, and consistent accuracy. The procedure can be used for full edge-shaping and trimming, surface grooving, and recessed carving.

When a template is used for full edge trimming, the workpieces need only to be rough cut and made slightly oversize. A straight bit of the same diameter as the guide pin results in a smoothly formed edge that perfectly matches the shape of the template.

The use of the template need not be reserved only for repeat work because its use is advantageous for one-time workpieces as well. Here is why: it is easier to cut and smooth a template made of $^1/_4$-inch hardboard or plywood than it would be to do the same with a $^3/_4$-inch-thick piece of hardwood, for example. Also, an error made in the template would be less costly than one made in the work.

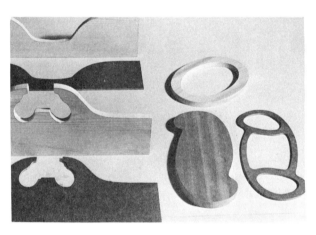

These are the workpieces and templates used in the projects illustrated in this section.

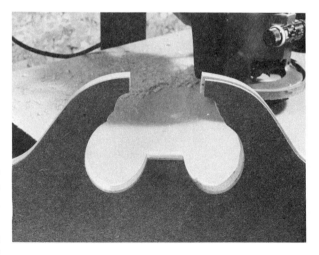

This shows the result of trimming a rough edge to size and shape with a pin-guided template and a straight bit. The section on the left has been trimmed, the right is still rough.

Straight or irregular grooves can be cut in the surface of the work by attaching a template cut to the size of the inner edge of the desired groove. Or the template may have one or more inside cutouts, in which case the dimensions would relate to the outside edge of the desired groove or grooves.

Two examples of the use of templates are shown: the fabrication of a decorative pediment and the routing of a tray. The pediment is made with dual templates; the tray utilizes a single, inside cutout template.

To make the pediment, one template is cut to the shape of the back panel; the second one is contoured to match the shape of the inside edge of the molded trim. The respective boards are then rough cut to size, allowing a bit of overhang. The templates are then nailed to the undersides of the work. A straight trimming cut is made to shape and

smooth finish the top edge of the back panel. A shaped bit is used to mold the inside edge of the trim piece.

The trim pieces are cut out of the board and glued onto the back panel with the top edges slightly overhanging. A final trimming cut with a straight bit flushes both top edges.

The tray is made by tack nailing the template hardboard stock to the workpiece. Both are cut together to form the outside shape; they are then sanded. The template piece is temporarily removed to saw the inside cutouts, and it is then reinstated by nailing through the same holes to obtain exact alignment.

Using a core box bit, the depth stop is adjusted for a cut about $1/8$ inch deep. The work is positioned with one of the cutouts over the guide pin. A plunge cut is made, and the carriage is locked. Then

Making the final pass to shape the edge of the molded trim. A pass with a beading bit, followed by a pass with a cove bit, produced the desired profile.

The sawed and sectioned trim pieces are glued to the back panel. Tempered hardboard templates like the one by the router can be used many times over.

A straight bit is used to make the trimming cut of the overhanging top edge.

against the pin to groove the perimeter of the recess. This is repeated in the other sections. The work is then shifted about freehand to rout the surfaces between the bordering grooves. The recessing is continued by taking gradually deeper cuts in increments of about $1/8$ inch until the desired depth is reached. A corner-rounding bit is used to ease the edges by guiding both the inside and outside edges against the pin.

The picture frame is partially edge-shaped and rabbeted using ball-bearing, self-piloted bits; therefore, neither a guide pin nor a template is required.

3. Freehand back-and-forth movements clear out the waste. The template prevents overshooting the edge.

1. Bottom view of the workpiece for the tray with the template attached for carving.

4. Router is adjusted for very shallow bites to make the final leveling cuts in the tray.

2. Carving is started with perimeter groove cuts made by guiding the template edges against the pin.

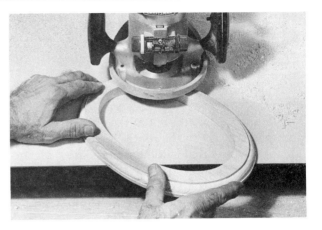

Making a beading cut to shape the edge of the oval frame.

Guide pin is withdrawn when the work is to be guided by a ball-bearing, self-piloted bit.

Piloted rabbeting bit is used to cut the recess for the picture in the back of the frame.

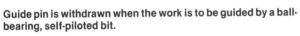

A clamped fence is used to guide the work for making mortise cuts.

Routing Hint

An important rule to observe when routing wide areas that are recessed to an edge (such as an end half-lap) is to start the cut at the end of the work and gradually move the router toward the inside. This will ensure that half the router base is always firmly supported. Working the router in the opposite direction would gradually diminish the support, eventually causing the base to drop off the ledge and mar the work.

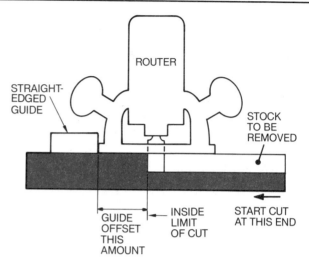

A wide lap cut should be started from the outside edge so that the router base is firmly supported during the entire operation.

Avoid Splintered Dadoes

On occasion disastrous splintering will occur along the edges of cross-grain dado cuts in veneer plywood. More often than not this will happen on the back, rather than on the face, veneer, and it is due to the extremely dry wood used for the lesser quality back veneer. The problem can be avoided by scoring the borders of the groove using a straight edge and a sharp knife before making the cut. Make the scoring cuts a fraction *outside* the lines of the cut and the groove should cut clean.

Compare the difference in these cuts made through the back veneer of ash plywood. The dado on the right was pre-scored.

Shaping Before Assembly

Constructions that include shaped or molded edges fall into two basic categories: those that can be shaped after assembly and those in which the shaping operation must be carried out in stages prior to assembly. Careful consideration should be given to this matter before assembly to avoid the problem of limited access to all the parts once they have been assembled.

The table construction shown is a typical example of a project requiring advance shaping of various components before final assembly. Notice that if the legs were attached before shaping the rounds on the inside, corners could be only partially shaped because the rails would obstruct the router base and prevent a full pass. The same would apply to the lower edges of the rails; the router would be stopped by the legs several inches short of the ends. Ease of working is another factor in favor of shaping before assembly when feasible. Consider how difficult it would have been to manipulate the assembled table on the disc sander to shape the leg tops.

The treatment shown will serve as a guide because the basics apply in practically any home workshop construction.

Lower edges of the rails require stopped rounds to the points where the legs join. A clamped block serves as a stop for the piloted bit so it will not overshoot the mark.

The router has full access to all the corners of the legs while they are still unattached.

Legs are temporarily clamped in place to serve as stops while rounding the top edges of the rails.

Final glue-assembly takes place after all the essential shaping work has been done.

Legs are removed to permit controlled contouring of the ends on a sander. This would be rather difficult to do after assembly.

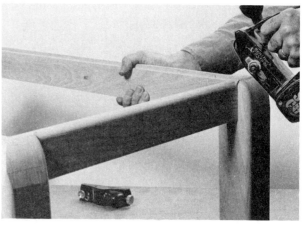

Freehand power sanding completes the shaping of the legs. The router-shaped edges will require only finish sanding.

16. SANDING

Jointer - Sander for Long Edges

While it looks like an ordinary disc-sanding wheel, the jointer-sander is quite different in shape and function.

An ordinary wheel is suitable for sanding small flats that fit within its radius, but it cannot be used effectively for sanding long edges for several reasons. Assuming precision sanding is desired, passing the work between the disc and a guide is not feasible because the guide would have to be parallel with the disc surface. This would not allow lead-in space and the work would be stopped by the rim of the wheel. The alternative would be to pass the work through freehand. This cannot produce precise results particularly because the work will pass from the proper, "down" side of the rotation to the wrong, "up" side. At best the edge would be erratic and show curved sanding marks.

The jointer-sander wheel avoids these problems and permits controlled straight-line sanding with excellent results. This is possible because the wheel is not flat but conical in shape, angled at 2 degrees. It is designed for use on a table or radial-arm saw. In use, the arbor is tilted 2 degrees. This presents a central, vertical point of contact between the disc and the work. Thus the forward and rear portions

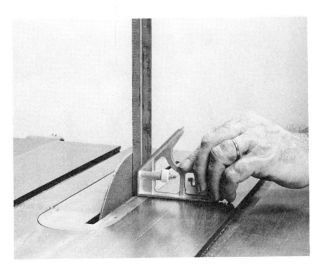

Saw arbor is tilted so the surface of the disc at center will be perfectly perpendicular to the table.

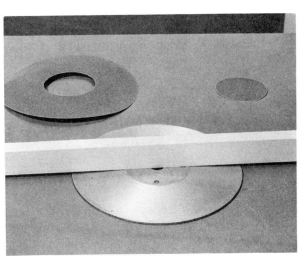

This shows the 2-degree conical slant of the disc. The abrasive disc available for this accessory has a center cutout.

173

of the disc curve away from the work, allowing it to pass smoothly through being abraided only by relatively horizontal motion. With a fence guiding the work, the result is a true, straight surface virtually free of sanding marks. The smoothness of the sanded edge equals that produced by a finishing sander.

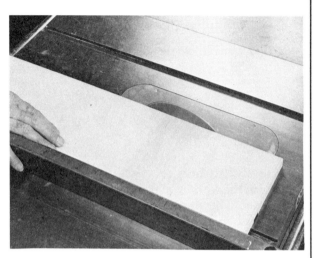

Guided by the fence, the work passes through without making contact at the edges of the disc. The fence should be adjusted for a maximum shaving of about 1/32 inch. The accessory is not designed for heavy stock removal. However, repeated passes with fence shifts can remove an appreciable amount of stock. The flat back of the disc can be used for conventional free-hand sanding of small flats, angles, and curves.

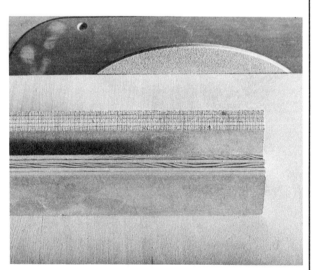

Comparison of edges sanded with conventional flat disc (above) and jointer-sander disc (below). Both were sanded with relatively coarse 80-grit abrasive.

Avoiding Splintering

When sanding wood in the direction of the grain, as it always should be done, never lead with the exposed edge of the abrasive sheet. To do so could tear out a splinter when the edge of the paper strikes a loose sliver of grain and digs in. Shown here is what frequently happens with fir plywood in particular when the wrong method is used. The problem is easily avoided by leading with the upturned edge of the abrasive sheet. The same rule applies when using power sanders.

Stopped Sanding

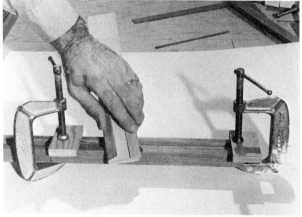

When it is necessary to ease only a part of a sharp corner, simply clamp stop blocks at the appropriate locations to limit the sanding strokes. This arrangement can be used for power sanding as well.

Gang Sanding Edges

Stack similar pieces together to save time and energy in sanding edges. The broader contact surface also makes it easier to obtain true flats because the sander won't have the tendency to rock as it usually does when working on a narrow edge.

Neater Butt Joints

For a neater job with right-angle butt joints, cut the add-on parts a fraction oversize so they overhang slightly. Use a belt sander to flush-trim the edges to the surfaces.

Jointing with a Sanding Drum

An edge can also be jointed true and smooth by passing it between a sanding drum and a guide fence. Adjust the fence so only a small amount of stock is removed with each pass-through. Feed against the rotation of the drum.

Dual Grit Disc

If you have only one disc sander, as is usually the case in home workshops, rough and follow-up fine sanding becomes somewhat of a nuisance because it means switching abrasives. This is not only time-consuming but costly, because the tem-porarily removed abrasive infrequently sticks well when reapplied. A novel solution is to cut out the center portion of an abrasive disc and insert a disc of a different grit in the cutout. This will let you enjoy the convenience of two grits on the same wheel. To make the inlays, place both abrasive discs on a board in exact outside alignment. Then trace around a circular object—a can will do—with a sharp utility knife. *Do not* discard the cutoffs—they can be used in reverse order the next time replacement is necessary.

Scroll Saw Sander

A scroll saw sanding accessory is useful for truing and smoothing small edges after sawing. The fixture accepts a ³⁄₄-inch-diameter sleeve and forms it half-round on one side and flat on the other so it can be used for both flat and radial edges.

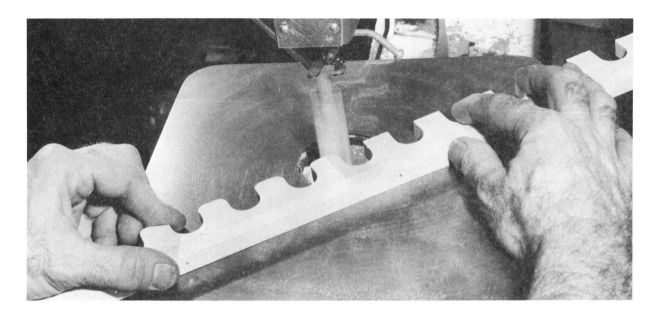

Flexible Strip

A contour such as this is best sanded with a strip of cloth-backed abrasive. Soften the abrasive first by kneading it between the fingers so it will conform to the shape you are sanding without wrinkling.

Custom - Sanding Sticks

Sand irregular edges better with homemade shaped sanding sticks. Make up a variety of cross sections using softwood and readymade moldings.

Use staples to attach the abrasive sheets so they can easily be replaced. However, curved cove molding requires a dab of rubber cement to hold the paper in the hollow.

Using a stick on an inside curve.

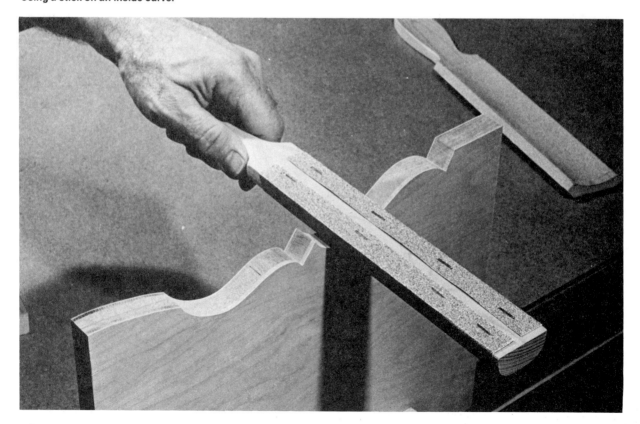

Sanding Cylinder

Use dowels to make small-diameter sanding cylinders for use in the drill press. Join the paper to the dowel with a strip of tape, then wrap tape around the ends to keep it from unfurling.

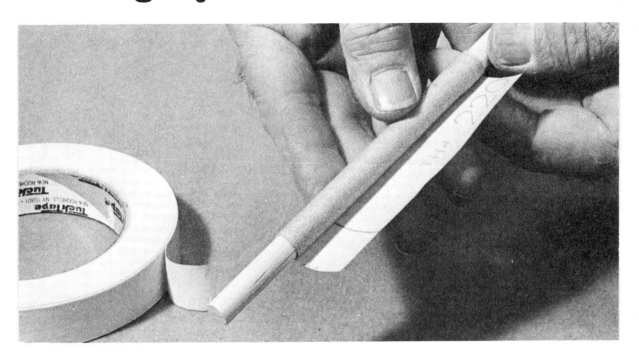

Sanding Hint

To obtain a good fit in a box and lid with rounded corners, tack the lid to the box with brads and sand them together.

Use a knife or thin scraper to pry the lid apart. Then remove the brads and fill the holes.

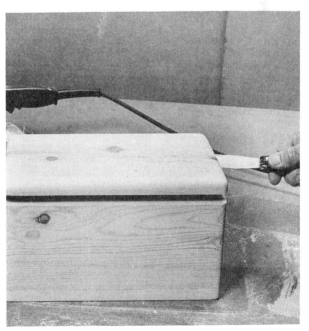

Cut Sanding Costs

Abrasive papers are relatively costly particularly in view of the fact that rapid wear seems to be a built-in characteristic of many brands on the market. You can cut down on costs by using substitutes for preliminary stock removal and smoothing operations. Two possibilities are shown. Others include the Surform tools and "sanding" sheets composed of tough carbide grits embedded in thin metal sheet which can be used in pad finishing sanders.

A cabinet scraper made of fine, flexible steel is a relatively inexpensive tool that will outlast sandpaper many times over for smoothing wood. It needs only to be followed by light sanding with extra-fine paper for most finish-ready applications. A sharp hook on the square edge does the cutting as the tool is pulled along the surface at an angle. The hook is easily sharpened with a hardened steel burnishing tool.

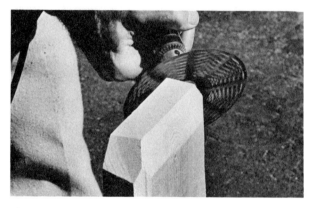

A rotary disc rasp chucked in a portable drill cuts aggressively and is useful for heavy stock removal. Due to the open tooth design the disc hardly ever clogs.

Sanding Irregular Shapes

A flexible sanding wheel is a handy accessory for use in the drill press, lathe, or portable drill. Thin strips of abrasive backed by brushes can work their way into most irregular contours to do their sanding work.

Sander / Grinder Techniques

The difference between the narrow-belt sander and the broad-belt portable and stationary sanders is more than size alone—it is versatility of function. The narrow belt can be used on wood, metal, plastics, and other materials to perform a wide variety of tasks, many of which cannot be done with the other sanders.

The narrow belt permits sanding or grinding in confined areas and, by threading the belt through openings in the work, internal surfaces can be shaped and smoothed. As a grinder, it is generally preferred over a conventional grinding wheel because it cuts faster and operates cooler. The latter is an important factor because it minimizes the danger of drawing the temper when sharpening tools or grinding metal in general. Some applications and techniques are shown; you can adapt many more.

A standard flat platen backs the belt for straight or outside-curve sanding. Saw ripples in this piece are quickly removed with a medium-grit belt.

A curved platen is substituted for sanding or grinding concave surfaces. For machines that do not have this accessory, a shaped block of wood can be clamped to the table directly behind the belt to serve as a platen.

Very tight spots can be handled by splitting the belt to obtain a strip of the required size. Belt is turned inside out and a small slit is made with a knife. Belt is then torn, a few inches at a time, alternately from both sides. This minimizes unraveling along the edges.

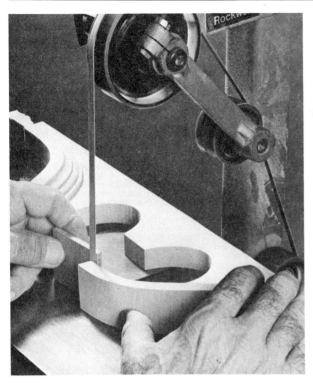

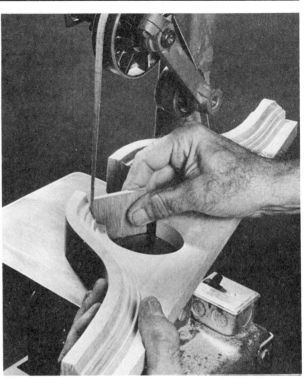

With the platen removed, a small block of wood is used as a back-up to guide the narrow belt into a corner.

Block is shifted to steer the belt over an endwise curved bead. Back-up blocks can be shaped for varied contours.

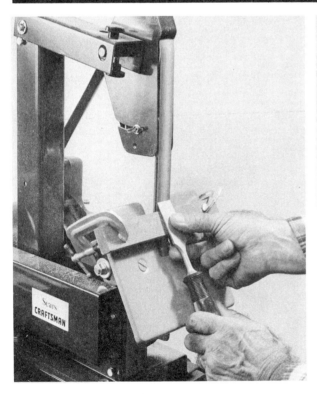

This is a good set-up for sharpening chisels and plane irons. The wood support is notched at the back by sanding, then it is clamped so there is a gap of about 1/16 inch between it and the belt. The back-up platen must *always* be used for sharpening.

Low-melting-point grease applied to the belt will prevent it from "loading" with metal grit when grinding aluminum and other nonferrous metals.

Accessories available for the Rockwell machine include abrasive web brushing/cleaning belt, nonabrasive belt, stick rouge, convex and narrow platens (shown with contact wheel).

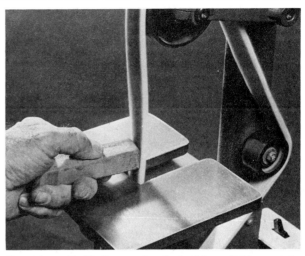

Rouge is applied to the nonabrasive belt for polishing metals and plastics.

The contact wheel has a rubber tire that is grooved to dissipate heat from the belt.

The nonabrasive belt excels for buffing acrylic plastic edges to a luminous high polish. Table and platen are removed for this operation.

The contact wheel itself is used with a coarse-grit belt for metalwork chores such as heavy stock removal, deburring, and descaling.

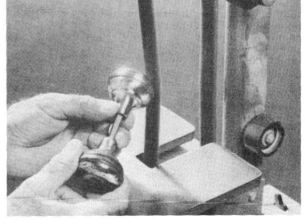

The brushing/cleaning belt is used for removing tarnish or rust on metal and for cleaning glass and plastics. The belt is made of nylon web impregnated with abrasive.

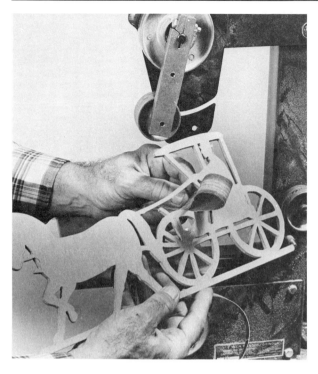

Belt can be threaded through an opening to permit internal sanding on complicated cuts.

Slightly-used fine-grit belt is best for work such as this because it has more flex than a new one. The platen is removed for this operation.

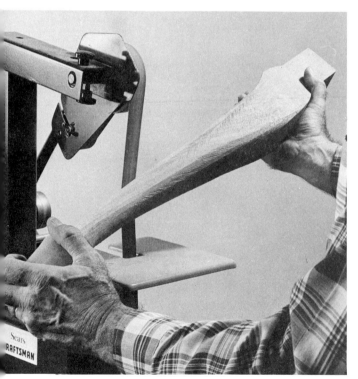

Coarse-grit belt does quick work in sanding out rough tool marks in wood, but care must be exercised in steering the work to prevent the belt edge from slicing into the surface.

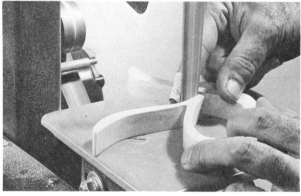

Pad the index finger with a Band-Aid to keep it cool while backing the belt to conform to irregular curves. This method provides good control.

With the platen removed, the belt's thin edge can be used to advantage in fitting tight spots.

17. ASSEMBLY

Spreading Glue on Edges

Another method for applying glue: run a wavy bead on one of the contact surfaces only.

Slide both members together to transfer the glue to the second surface.

Spreading Glue on Surfaces

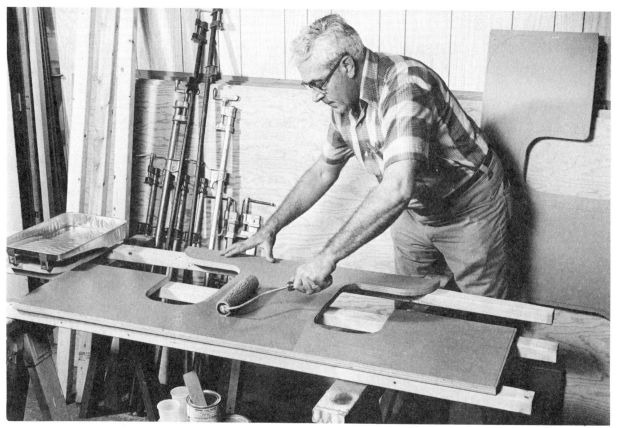

Use a short-nap paint roller to spread glue on large surfaces. This method is so quick that it permits the convenience of using glues with relatively short assembly times. The uniform coating avoids messy excessive squeeze-out and waste.

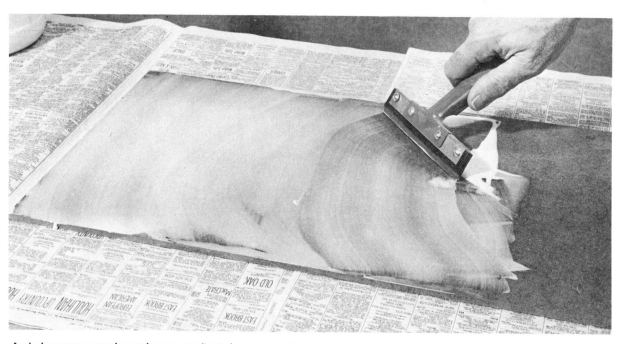

A window squeegee also makes an excellent glue spreader for large surfaces.

Assembling Miters

Hot-melt glue is also effective for assembling mitered frames. The glue sets in 15 seconds, so you can simply hold the joints together. The assembly is done in three steps. Start by joining one corner to form an L. Repeat this with the corner diagonally opposite. When the two L's have been formed, apply glue to two ends and bring the two sections together. Use waxed paper under the joints to prevent the work from sticking to the table. When the glue has set, reinforce the joints by driving two nails into each corner at right angles to each other.

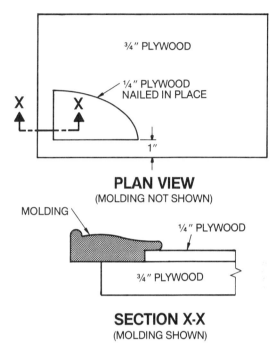

³⁄₄" PLYWOOD

¹⁄₄" PLYWOOD NAILED IN PLACE

1"

PLAN VIEW
(MOLDING NOT SHOWN)

MOLDING

¹⁄₄" PLYWOOD

³⁄₄" PLYWOOD

SECTION X-X
(MOLDING SHOWN)

Mitered picture frame corners are usually tricky to assemble, but the nail gun and this simple jig will make easy work of it. Position two molding strips with the rabbets butted against the corner guide, then clamp one strip to the backboard as shown, using a pad to protect the surface of the strip. Apply a dab of glue to both faces of the miter, hold the backboard firmly, and drive two nails into the corner. Reposition the frame and repeat the procedure.

A spline provides good reinforcement for a miter joint and helps to hold the parts in alignment during gluing. The spline can be made of solid wood or plywood, but in either case the grain direction must run at right angles to the face of the miter. It should be noted that in thin, three-layer plywood the greater strength is in the grain direction of the outer plies.

Tack nailed wood strips can be used in this manner to hold a glued mitered frame together.

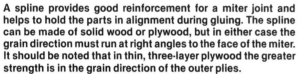

This is an emergency method for closing a stubborn outside corner of a miter.

Another method to assemble a mitered frame utilizes six clamps. The two end members are clamped firmly to the table. Two bar clamps are used in turn to clamp the sides to the ends. This will ensure a flat assembly.

Assembling Irregular Joints

Some assemblies present problems in gluing because of the difficulty or impossibility of attaching clamps to hold the joints together. This odd-shaped structure is a good example. There was no practical way to apply clamps when gluing the five-sided frame members together—so none were used.

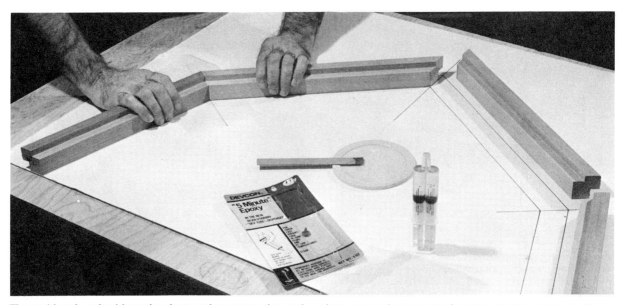

The problem is solved by using fast-setting epoxy glue and simply holding the joints together. Each joint face is coated, then pressed together on a flat surface for 5 minutes to form a strong, permanent, and waterproof bond. Unlike conventional glues, epoxy does not require great squeeze pressure; therefore moderate hand pressure is ample. Notice that waxed paper is used under the glue line to keep the work from sticking to the work surface.

Another example of an assembly that does not lend itself to clamping. This shows how tape can be used to solve the problem quickly and effectively.

Further along in the assembly, wide duct tape is used to hold the glued part together. Although clamps could have been used at this stage, the work would have required a large number of them and much fuss and bother to apply them. The tape proved quick, easy, and effective.

Nail Assembly

Large framing square temporarily screwed to the worktable makes a good backup to hold small work in place and in true alignment while gluing and nailing.

Brad pusher does a neat quick job of driving and setting brads up to 1 inch long. A magnetic drive rod in the barrel lifts and holds the nail, which is then retracted into the barrel. A push on the handle drives it completely into the work.

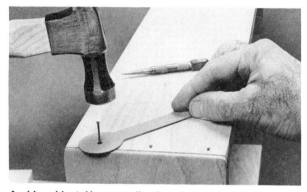

Avoid accidental hammer dimples when driving finishing nails on fine work by using a stop pad cut from a piece of 1/16-inch-thick plastic laminate. Place it over the started nail and drive the nail until the hammer strikes the pad lightly. This will leave the nail heads projecting 1/16 inch. Follow with a nail set to sink the nail below the surface.

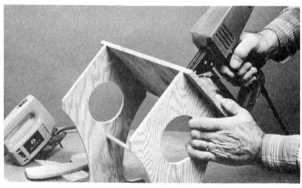

If driving nails into thin edges with a hammer proves troublesome for you, try using a nail gun. It drives nails remarkably straight; and since the nail is driven instantaneously, there is no bouncing as occurs with hammering. Thus the danger of the parts shifting out of alignment is avoided. Also, the blunt point of the nail prevents splitting even in very narrow stock.

Nail-spinner drill accessory is a handy gadget for driving finishing nails up to 1½ inches long without the danger of splitting the wood. It slips over the head of the nail and is easily released by pulling it away when the nail is almost driven.

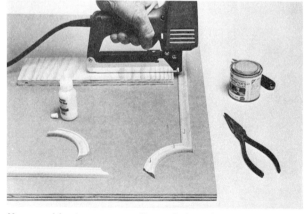

You can drive temporary nails partly into the work by placing the nail gun on a stand-off. In this application the nails are driven to hold the parts in place until the glue sets.

Before nail-and-glue-assembly of large panels, bore—from the inside and centered along the joints—pilot holes of slightly smaller diameter than the nails. This will enable you to spot the nailing locations readily from the outside and will ensure that the nails will enter the second members centered.

Use temporary cleats to obtain good alignment of the parts when nailing. Drive the cleat nails only part way in so they can be removed readily. If glue is used in the assembly, make sure to remove the cleats as soon as possible so that any excess glue will not stick them to the work.

Installing Hinges

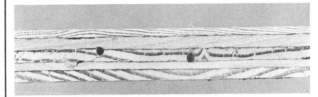

Drilling pilot holes perfectly centered into the end grain of the middle ply of fir plywood when installing a strip hinge can be a very frustrating experience. If a hole coincides with a dense summerwood annual ring, the drill bit will invariably be deflected off course and into the softer springwood alongside, as shown in this close-up.

Avoid the problem by first tack nailing a 1/8-by-3/4-inch strip of solid wood on the edge to be drilled. Use small 1/2- or 5/8-inch nails so the strip can be removed easily. Drive the nails about 6 inches apart so the strip stays firmly in place.

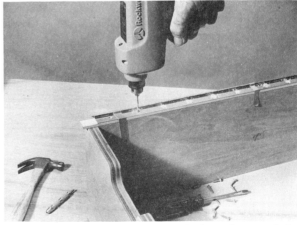

Align the hinge on the strip and tape it in place, then punch the center marks and drill the holes. The drill bit will not slide off course and damage the work.

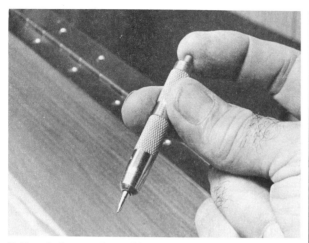

Self-centering punch provides a simple method for accurately centermarking for screw holes when attaching hinges or any hardware that has countersunk holes for screws. The plunger is spring-loaded and retracts when at rest.

The tapered end of the punch automatically centers when it is placed into the countersunk hole. A light tap on the plunger makes the mark.

T and strap hinges in particular have quite a bit of play, or looseness, between the leaves, and this could result in a poorly hung door unless you go about the installation in the proper manner. The procedure is simple. Insert shims at the sides and bottom of the door to hold it in centered alignment. Attach the jamb leaf first, then press down on the door leaf while marking the screw-hole centers. This will compensate for the play and the door will hang true.

Dovetail Corner Block

Add strength and rigidity to rail and leg constructions with a dovetail/dado corner block such as this. Cuts are easily made on the table or radial-arm saw. A lag bolt inserted through the center of the block and into the corner of the leg will further reinforce the assembly.

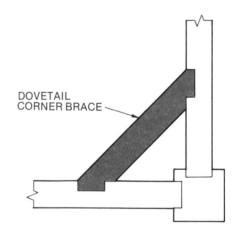

DOVETAIL CORNER BRACE

Glue Injector

Use a small-nozzle glue injector to apply glue neatly in hard-to-reach or confined areas. Available from woodworker's mail order supply outlets, the accessory is particularly useful for fine repair work.

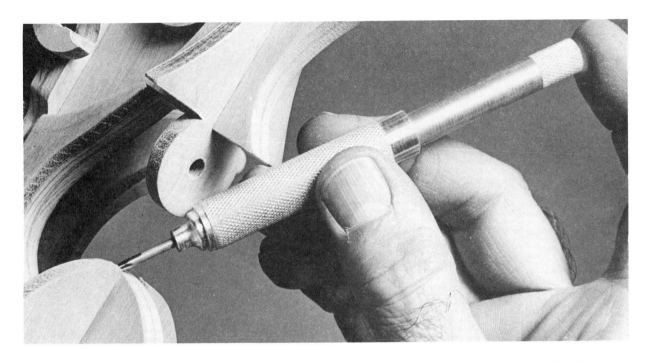

Drawer Front Alignment

On projects that involve drawers with add-on fronts, you can be assured of obtaining perfectly centered fronts if you follow these steps. Construct the drawers and fronts, but do not join them until the drawer-slide hardware has been installed. Insert the drawers in place, then drive two nails partly through the sub-front so the points protrude about 1/8 inch. Back off the drawer slightly, then place the front panel in position; use shims to hold it centered. While supporting the front with one hand, slowly move the drawer forward until the nail points make contact. Remove the drawer and attach the front permanently using the nail points and the corresponding registration holes for positive alignment. The procedure is reversed when there is no open access to the drawer. In such a case, block the drawer so it will not move back and prop the front in the desired position and press it forward to meet the drawer.

Long Clamp Substitute

When the workpiece is longer than the capacity of the available clamps, you can usually solve the problem with a cheater such as the one used here made up of hand screws, scrap wood, and bar clamps.

Reinforcing with Gussets

Mitered and butt framing joints can be strengthened with the use of double plywood gussets, nailed and glued.

Hot-Melt Glue

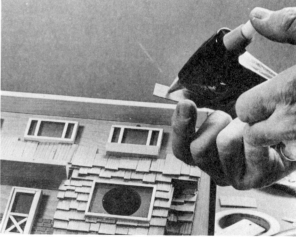

Hot-melt glue is a practical way to assemble numerous small parts. It is easily applied and sets so quickly there is no time wasted in waiting for glue to dry.

Bolt Assembly

To assemble a dado joint with a bolt, bore a perpendicular hole slightly deeper than the length of the bolt.

Bore the cross hole to intersect the bolt hole. Bore slightly beyond the bolt hole to allow clearance for the nut and washer. A tape marker on the drill bit is used to gauge the depth.

Carefully mark the cross hole for the nut so it will be centered over the end of the bolt.

Balance the nut and washer on your finger and slip it into the hole to reach the bolt end.

18. MISCELLANEOUS

Marking Softwood

When coding parts made of very soft wood, such as pine or redwood, try using chalk instead of pencil. Pencil marks indent these woods and require tedious sanding to remove. The chalk will not mar the surface, and the marks are easily removed with a damp cloth.

End Center Marker

Use this jig to mark centers accurately on the ends of round or square stock up to 4 inches across. Make the support cradle by cutting a 45-degree bevel in two pieces of 2 × 4 stock. The triangle must be carefully positioned so its edge bisects the center of the V.

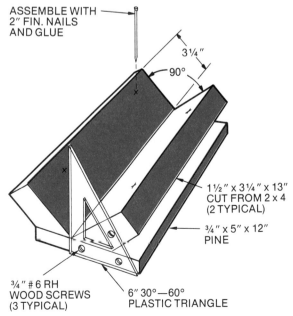

ASSEMBLE WITH 2" FIN. NAILS AND GLUE

3¼"

90°

1½" x 3¼" x 13" CUT FROM 2 x 4 (2 TYPICAL)

¾" x 5" x 12" PINE

¾" # 6 RH WOOD SCREWS (3 TYPICAL)

6" 30°—60° PLASTIC TRIANGLE

Board Center Marker

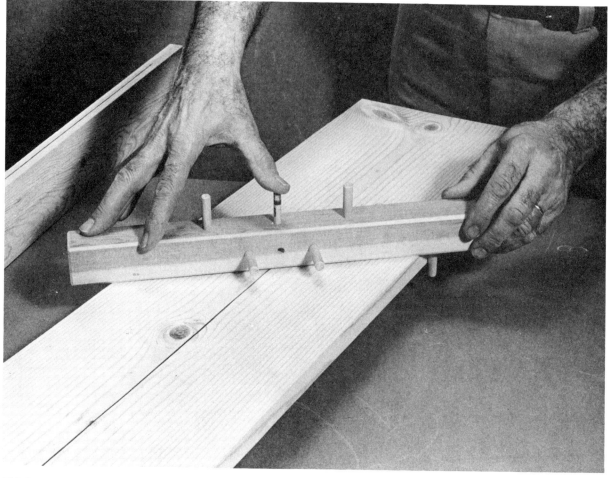

This handy jig marks a line down the center of boards of varying widths. It is placed on the stock with the projecting dowel pins butted against the edges and is pulled back while applying light pressure on the pencil stub. Dowels are variously spaced on each face in order to suit boards of different widths. Face with the closely spaced dowels is used to mark edges.

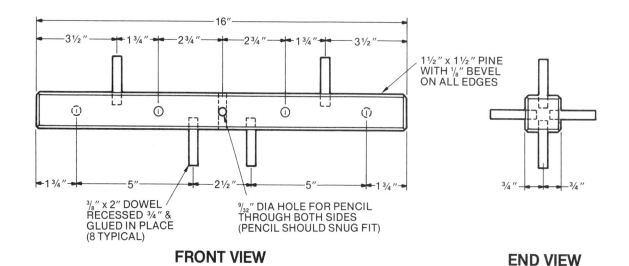

FRONT VIEW

END VIEW

Tracing Wheel

A dressmaker's tracing wheel is handy for transferring irregular outlines from drawings to wood. Insert carbon paper under the drawing and roll the serrated wheel over the lines. The result will be a dotted pattern, much smoother than would be obtained by tracing with a pencil.

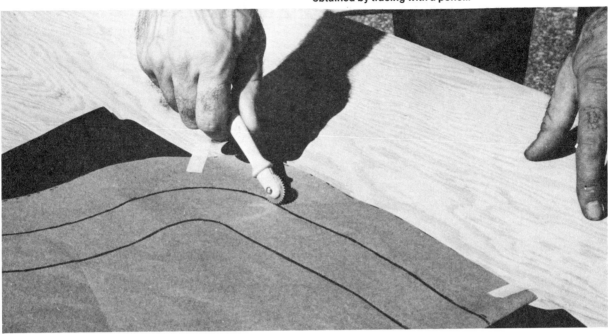

Painting Mask

To mask a surface for painting a second color, use mylar packaging tape instead of regular masking tape. The thin, smooth mylar makes closer contact so it minimizes the danger of the paint running under. Use a compass with a knife-blade acces-sory (substituted for the pencil lead) to cut the tape parallel to an irregular edge. Then strip the tape from the area to be painted. After painting, peel the remaining tape off for a neat outline around the edge.

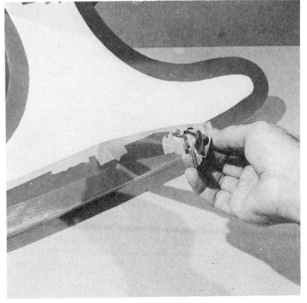

Laminate Shop Charts

Use clear, self-adhesive plastic to preserve shop data sheets. Cut the material slightly oversize and apply to both sides. Trim with a razor blade so about ⅛ inch overhangs the edges.

Rub-On Type

Use readymade press-type, which is transferred by rubbing, to obtain a neat, professional look on projects requiring numbers or lettering. Striping tape, available in various widths, is also useful for adding a finishing touch. Both items can be obtained at art supply or stationery stores.

Locate Framing

Before you conceal wall or ceiling framing permanently in home remodeling work, make a positive record of the exact location of the framing members. This may prove very useful in the future when additional work is to be done. Studs and joists are supposed to be evenly spaced at standard intervals, but frequently they are not. Use a roll of adding-machine paper to mark the location of each framing member, identify the roll, and file it where you can find it.

Saw Safety

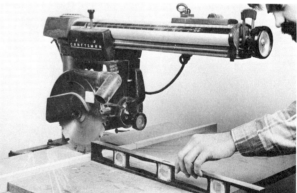

Radial-arm saws on mobile bases are advantageous because they can be moved about for use in different locations when the need arises. But you must be careful when you temporarily shift to a new location. If the floor pitches downward toward the front of the saw, so will the arm of the saw. This is potentially dangerous, because upon completion of a cut the still-rotating blade could cause the carriage to move downhill toward your unsuspecting hands. Check with a level and use shims on the floor to level the saw if necessary.

Cardboard Models

When a project involves tricky shapes or assemblies it is a good idea to make the parts out of cardboard so they can be checked in advance for size and fit.

Use the cardboard as templates to mark the work after any required corrections have been made.

Fly Cutter Safety

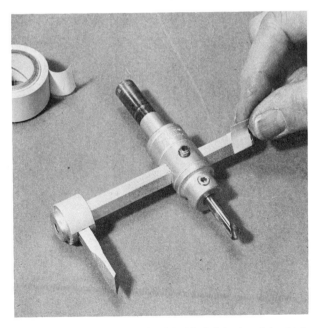

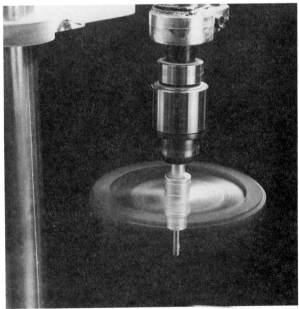

Fly cutters are usually made of darkly finished metal, which results in poor visibility of the tool as it spins in the drill press. This could prove quite dangerous to a misplaced hand. Make the tool highly visible by spray-painting it with aluminum paint. Mask off the shank, drill bit, and cutting edge so they do not get painted. For added brightness, apply fluorescent colored tape to the ends of the bar. The outline of the newly decorated fast-spinning cutter is easy to see.

Cutting Rings

When you need small wood rings, use two blades simultaneously in the hole saw. A single cut will produce a finished ring. The diameter and wall size can be varied within the range of the blade diameters.

Flexible Shaft

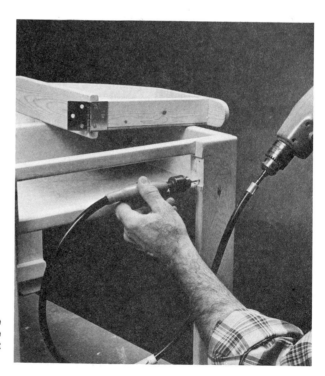

The flexible shaft is a very handy accessory for your portable drill as it will enable you to bore holes in close quarters where the drill housing would prevent access. Always hold the shaft in a wide loop to prevent it from twisting around itself.

Counterboring Trick

When it is necessary to bore a large through hole and counterbore hole, the normal procedure is to bore the larger hole first so the center for the smaller bit will be preserved. Since the center hole made by the larger bit will be too large to center the smaller bit properly the second hole will very likely have to be bored from the back. Unless a back-up is provided the bit will splinter the wood severely as it exits in the hole. Prevent this by inserting a roughly cut circular back-up block. Use strong-gripping duct tape to keep the block from spinning.

Aligning Dowel Holes

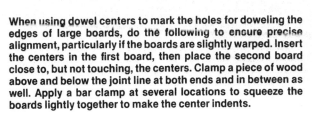

When using dowel centers to mark the holes for doweling the edges of large boards, do the following to ensure precise alignment, particularly if the boards are slightly warped. Insert the centers in the first board, then place the second board close to, but not touching, the centers. Clamp a piece of wood above and below the joint line at both ends and in between as well. Apply a bar clamp at several locations to squeeze the boards lightly together to make the center indents.

Cord Strain Relief

Although most portable power tools have built-in strain-relief sleeves where the power cord enters the housing, they are usually too short and stiff to prevent the wires from breaking from constant flexing. Improve the condition by adding a supplementary support made by twisting a piece of single-strand # 14 electrical wire around the cord as shown.

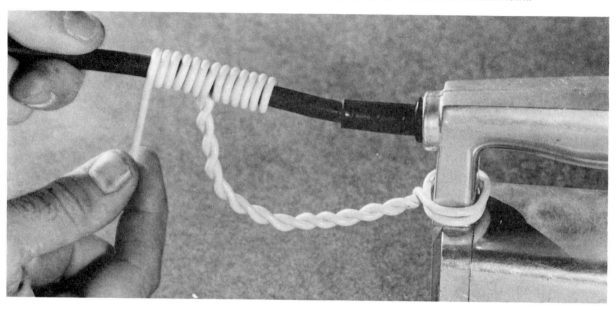

Sliding Tape

Save on the expense and bother of installing drawer-sliding hardware on cabinet drawers by utilizing self-adhesive sliding tape instead. An added advantage is that the tape permits designing drawers of maximum size since very little clearance space is required. The tough plastic tape is applied by removing the protective backing and pressing into place. It is available through most woodworker's supply outlets.

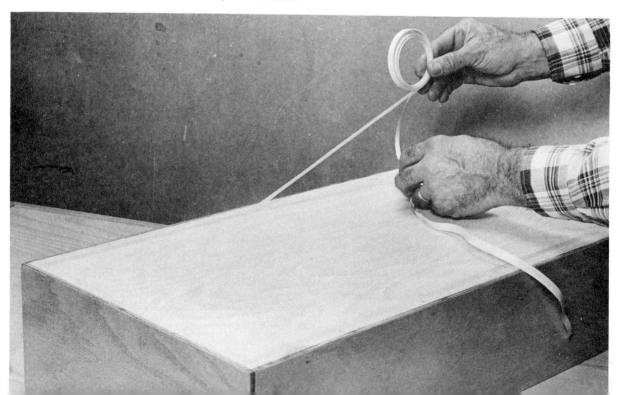

Cutting Foam

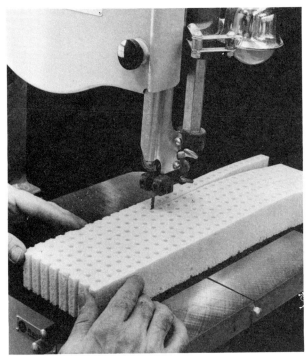

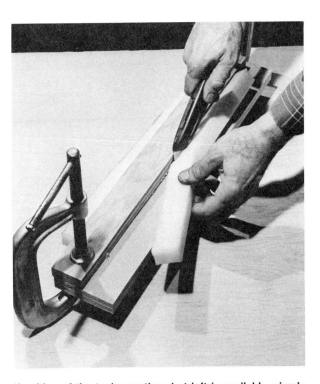

Spongy foam used for padding in chair and bench projects is rather difficult to cut accurately with scissors. For perfect results use the band saw.

If neither of the tools mentioned at left is available, simply clamp the foam between two pieces of wood to compress it as much as possible. Then make the cut with a sharp utility knife.

An electric carving knife also does a fine job cutting foam. Lay a board on the cutting line to guide the knife for a straight cut.

For irregular cuts, use a shaped board or a heavy cardboard template to guide the knife.

Wood Filler Patch

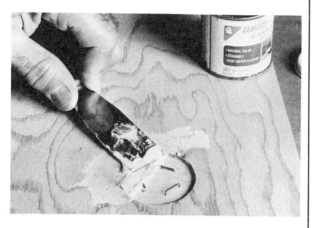

A large wood filler patch in wood will hold much better if it is "keyed" to lock it in place. Use a few partly driven staples for this. Tilt the stapler slightly into the hole without allowing its end to touch bottom. This will set the staples slightly below the surface.

Secure Carriage Bolts

In some applications where wing nuts are constantly adjusted on carriage bolts, the latter eventually work loose. Avoid the problem by counterboring slightly to recess the head of the bolt, then apply epoxy glue around the head to lock it in place. A few V notches filed into the edge of the head, as shown, will give added locking power.

Securing the Work

Here is a novel way to support work when a workbench is not available.

Scroll Saw Magnifier

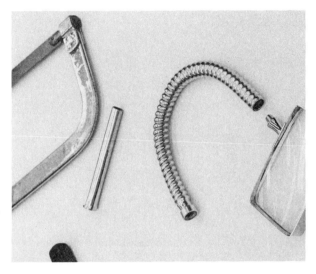

Get a close-up view of the cutting line for precision work on the scroll saw by attaching an adjustable magnifying glass. This one is made to mount in the auxiliary-blade-guide assembly bracket on the Rockwell saw. With a slight modification of the mounting method it can be adapted to any saw.

Arm is made from a length of flexible lavatory riser tube. Cut off the straight portion of this copper piping, remove the handle from a reading magnifying glass, and use epoxy putty to join the two.

Tool Table Care

Machine table surfaces should be kept spotless and free from corrosion if they are to function friction-free. Tables that have been neglected can be restored by cleaning with automotive rubbing compound followed with a light rubbing using the companion polishing compound. These are creamlike products made by DuPont and are sold at auto supply stores. Keep the cleaned surfaces smooth and slick with an occasional application of a lubricant such as WD-40.

Scroll Saw File

A regular file can be adapted for use in a scroll saw by grinding the tapered tang straight and to a size that will fit the lower chuck of the scroll saw. If the file is too long it can be shortened by wrapping it in cloth and securing it in a vise so the part to be removed projects above the jaws. Strike a sharp blow with a hammer to snap off the excess. Be sure to use the protective cloth because the brittle metal could splinter and pose a safety threat.

Blade Care

Circular-saw blades quite often cut poorly because they are dulled by gum and pitch accumulation. Get optimal performance from your blades by soaking them for a few minutes in a gum and pitch removing solution whenever they show signs of a build-up. The solution can be used over and over again so the cost is negligible.

Spare Costly Solvent

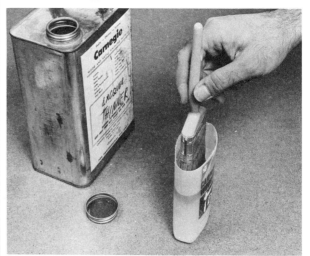

You can soak a paint brush properly, with all the bristles submerged, in a smaller volume of expensive solvent by using a narrow container. Cut the top from an empty flat-type plastic glue bottle. Collect a few sizes to accommodate brushes of varying sizes.

Dovetail Template

You can obtain perfect layout lines for dovetail joints by using a pair of matching templates to trace the lines. Making the templates will require extremely careful work; but once you have them, you will be able to lay out the cutting lines on the work with great speed and consistent accuracy.

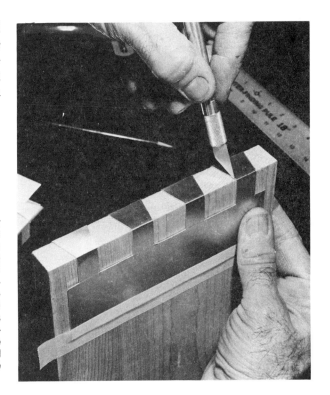

To make the pattern for the templates, fold plain writing paper over the edge of stock of the same thickness for which the template is to be used. Tape the paper firmly in place and use a square and T bevel to mark the lines. Make one drawing for the dovetails and one for the pins. Remove the drawings and mount them with rubber cement on a piece of light-gauge aluminum sheet. Use a sharp utility knife to cut through the aluminum by tracing along the lines with the aid of a metal straightedge. Work on a piece of hardwood so the metal does not deform downward as it is cut. In order to allow for bend-back loss, cut just outside the bottom lines. When the waste has been dropped out remove the paper pattern. Align and clamp the template to one side of the sample board, then use a stick of wood to form sharp corner bends over the edge.

Typical flat outline pattern for a dovetail template.

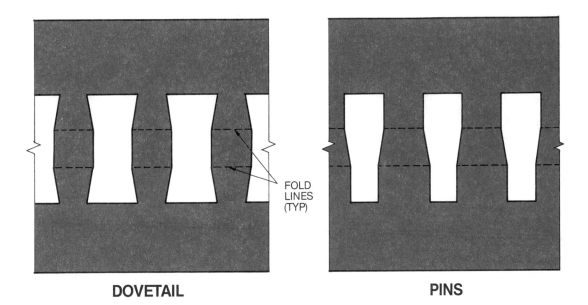

FOLD
LINES
(TYP)

DOVETAIL   PINS

Making a Tool Handle

ground back to the solid portion so the combined rivets will equal the thickness of the handle. A slight bevel is ground on the end of the shortened (male) piece for easier fitting. Follow the steps shown to make and fit a new handle.

Tools such as knives, spatulas, and the like are often discarded because of a worn or broken handle, but they need not be because a new replacement handle is easily made. It should be noted that regular hammered-over rivets are not used for handles of this type. Special knifemaker's rivets are generally used in manufacture, but these are not readily available. They feature deep heads with screwdriver slots. After threading the male into the female section, the projecting slotted portions are ground off flat to the handle surface. Lacking these special rivets you can nevertheless do a factory-quality job by substituting brass brake shoe rivets which are available at auto brake repair shops.

These rivets come in two sizes: # 5 and # 7. For this application you use one of each to make a set, because it so happens that the # 5 rivet will friction fit the hole of the # 7 rivet. The smaller rivet is

2. Insert a paper shim in the slot to keep the wood from springing while sawing the outline. Sanding follows to smooth out the saw ripples.

1. Trace the outline of the old handle onto a piece of close-grained hardwood. Clamp a guide fence on the band saw and make a partial edgewise cut for the tang. If the slot needs widening shift the fence and make another pass.

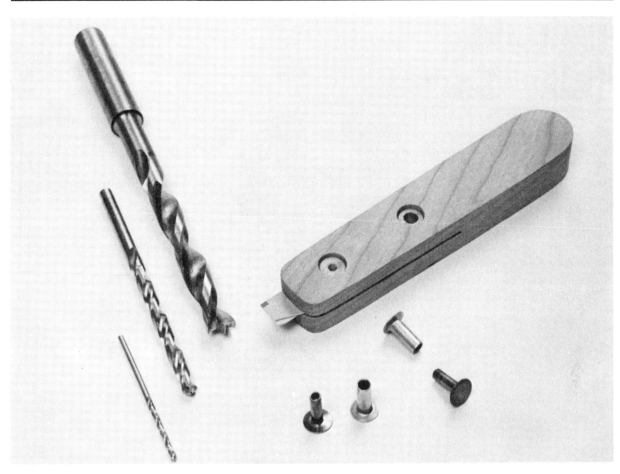

3. Three drill bits are needed to make the counterbored holes for the rivets. First, drill a ¹/₁₆-inch through hole to provide a centering hole on both sides of the handle. Next, drill a ³/₈-inch hole just deep enough to seat the rivet head flush to the sur- face. Use a brad point bit for this because it bores a flat- bottomed hole. Finally, rebore the through hole using a ¹³/₆₄-inch bit. The paper shim prevents flexing during drilling.

4. Round the edges on a disc or belt sander. To permit free movement, the work is not rested on the table.

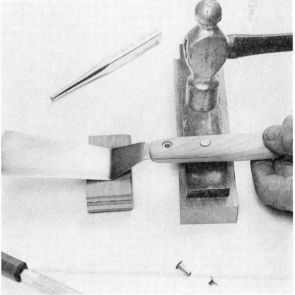

5. Force the rivets together by hammering over a metal block.

Imitation Inlay Turning

You can enjoy an interesting departure from conventional lathe work by making built-up turning blocks using contrasting woods. In the simplest form of this technique woods of varying species and thicknesses are assembled around a central core. More complex blocks can be made by cutting and regluing sub-assemblies. Designs emerge in different forms as the various layers of wood are cut through in shaping the contours. The design possibilities depend on the makeup of the block and the

Built-up block results in a turning apparently made up by intricate inlaying.

shape of the turning and are limited only by the imagination. Complex inlay effects can be achieved. But the results cannot be predicted in advance, and this makes the procedure all the more fascinating.

Try your hand at creating novel turning blocks. The procedure for making the block for the turning shown will serve as a guide for experimenting. It was built up with segments made by first laminating strips of mahogany and poplar side by side. These blanks were then cut into triangular pieces, which were in turn glued edge-to-edge to form squares with alternately offset bands of light and dark wood. The squares were then stack laminated, alternately exposing a light edge against a dark one. It should be noted that in order to obtain the triangular segments with alternately offset light and dark bands the initial lamination was composed of an equal number of light and dark strips. The process of cutting the miters by alternately flipping the blank up side down on the table saw produced pairs with the desired mismatched segments.

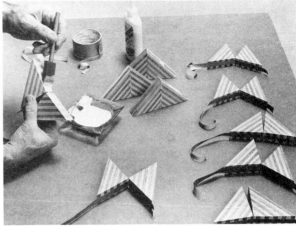

1. Striped blank is cut into triangular segments by making 45-degree miter cuts. Piece is flipped over after each cut. Pairs with light and dark outside edges result due to the equal number of light and dark strips in the blank.

2. Half-section segments are made up first. Strips of strong duct tape serve as clamps.

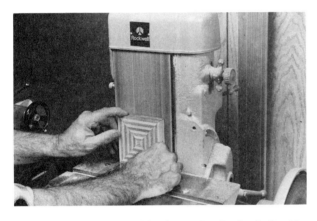

3. Surfaces that are to be joined must be absolutely flat. The belt sander or jointer can be used for this. If the jointer is used, it should be adjusted for very shallow cuts.

4. Segmented squares are stack laminated, one above the other, with alternate light and dark edges exposed. Finishing nails are driven in outside corners to keep the pieces in alignment during clamping. Areas with the nails are cut off by band sawing before turning.

5. Hardwood blocks are added to the ends to receive the lathe centers. Now the fun work begins.

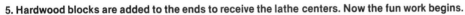

Auxiliary Vise Jaws

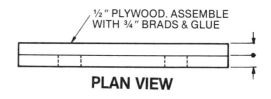

½" PLYWOOD. ASSEMBLE
WITH ¾" BRADS & GLUE

PLAN VIEW

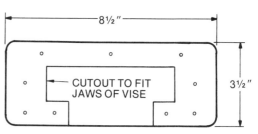

8½"

CUTOUT TO FIT
JAWS OF VISE

3½"

FRONT VIEW

These slip-on wood jaws permit the use of a metalworking vise for holding wood without the danger of marring the surface. The oversize pads also distribute the pressure over a greater area, thus hold the work better.

Applying a Plastic Laminate Self-Edge

Application of a self-edge (the same material on the edge as is used for the surface) to a counter is not especially difficult when a straight band is involved.

But difficulty can be experienced when the laminate strip must bend around a corner because a slight misalignment at the onset can cause the strip to go way off course after it turns the corner. This could ruin the work.

The normal procedure is to cut the laminate strip slightly oversize to permit flush-trimming to the surface, but this will not necessarily compensate for an appreciable misalignment. The best way to ensure good alignment of the edge strip to the counter edge is to shim the slab from below so it will stand off the worktable an amount equal to the amount of overhang desired. This procedure and the follow-up steps are shown.

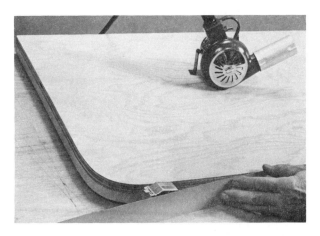

1. Place wood strips under the countertop so its edge will rest slightly above the surface of the worktable.

2. Insert paper in the gap to protect the table from adhesive drippings. Apply two coats of contact adhesive to the edges.

3. Position the laminate strip on the table, then bring it into contact with the edge starting at one end. Insert a barrier just before the curve to prevent premature contact.

4. Heat the laminate to permit bending around a relatively small radius without cracking. A flameless heat gun is ideal for this, but a heat lamp or a small hair dryer can be substituted. *Do not* overheat.

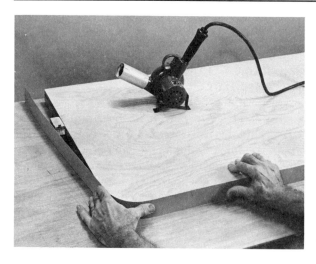
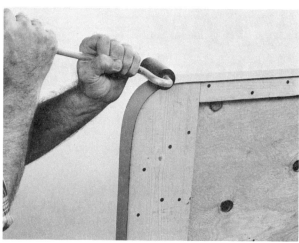

5. Shift the barrier as shown, then bend the heated strip around the corner and hold it in place while it cools. If you have tender fingertips wear a glove. When the strip has cooled for a minute or so, remove the barrier and continue the bond.

6. Use a roller to apply pressure over the bonded surface. If a roller is not available, hammer blows over a softwood block are an adequate substitute.

7. *Never* apply pressure to the overhang beyond the edge of the supporting core. To do so will invariably cause the laminate to split back into the good area.

8. Use a router with a flush-trimming straight bit to shave the overhang flush to the surface. Extra projection of the bit permits traversing the corner.

Adding the Laminate Top

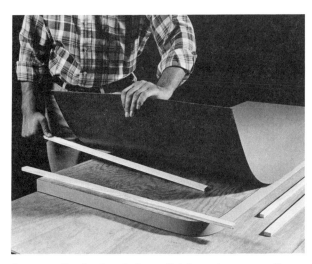

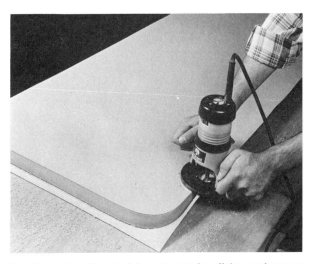

Contact-adhesive bonds immediately and permanently, so you do not get a second chance. Use slip sticks to ensure accurate alignment. Lay the sticks on the surface, position the laminate with overhang all around, then progressively pull out the sticks, one at a time, to allow contact to be made.

Use the router with a straight cutter to trim off the overhang as was done with the edge.

Finish off the edges with a bevel cutter. Adjust the depth of cut carefully and make a test cut on scrap first. Too deep a cut will expose wood at the corner and ruin the work.

Cutting Laminates

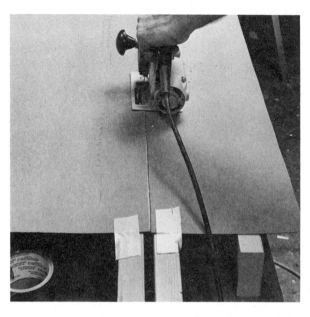

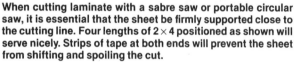

When cutting laminate with a sabre saw or portable circular saw, it is essential that the sheet be firmly supported close to the cutting line. Four lengths of 2 × 4 positioned as shown will serve nicely. Strips of tape at both ends will prevent the sheet from shifting and spoiling the cut.

Sheets of laminate which are to be butt joined edge-to-edge must match perfectly. The best way to accomplish this is with a router. Overlap the two sheets about an inch. Clamp both sheets to bottom supports that are spaced apart sufficiently to clear a straight cutter. Guide the router base with two parallel guide strips held in position with the same clamps. A cut through the center of the overlap will result in matching edges.

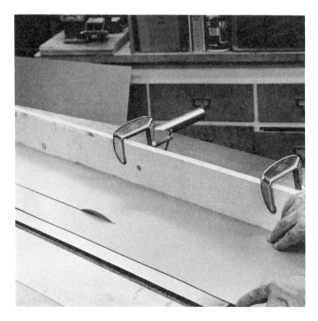

A fine-tooth plywood blade is ideal for cutting laminate. Clamp a board to the fence of the table saw to prevent the thin sheet from creeping into the gap between the table and the fence bottom. The decorative side of the sheet faces up with this saw. This also applies for the hand, scroll, band, and radial-arm saws. The decorative side should face down with a sabre or portable circular saw.

Laminate can be cut easily by scoring with a plastics scribing tool, then breaking. Place the sheet on a firm, flat surface, face side up. Tape a metal straightedge in place to guide the tool and make several firm passes. Turn the sheet face side down, then lift one side to snap it off.

Avoid saw blades with coarse teeth when cutting plastic laminate: this shows what usually will happen.

Plastic laminate is tough on sharp-edged cutting tools. Spare the edge of the cutting lip of a spur bit by pulling out the waste after the spur has just entered the wood. Then finish the cut.

Even smooth cutting saw blades will result in minute chipping. Always be sure to sand the sawed edge lightly to remove all loose particles before applying adhesive. Particles that get into the adhesive surface will prevent a good bond.

Cleaning Up Laminate Edges

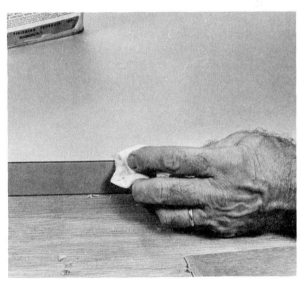

Use a scrap of laminate to scrape off any residual adhesive. This will not scratch the surface.

Wipe over smudges with a cloth dampened, *not* dripping, with lacquer thinner. Too much thinner could get into the joint and dissolve the adhesive.

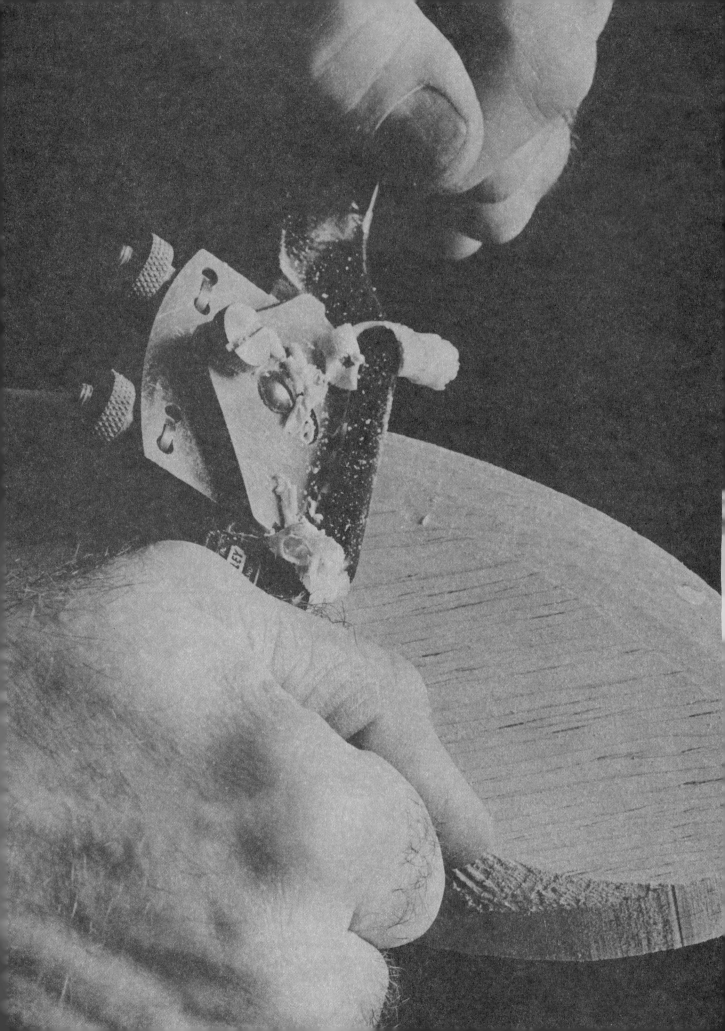

PART III

Projects

THE PROJECTS INCLUDED HERE differ widely in the techniques involved in their construction. They were selected on this basis in order to provide a hands-on approach for learning a variety of additional techniques. In the process of constructing a useful end-product you will acquire equally useful know-how. Should any of the projects fail to fulfill your specific needs or desires, simply study the step-by-step procedures and apply the basics to creative designs of your own.

19. THIRTY-SIX-SEGMENT REDWOOD PLANTER

If you have ever experienced difficulties in cutting and assembling tapered forms with a few segments, you have something in common with many woodworkers because this kind of work can indeed be challenging.

Thus you may be inclined to shy away from this project upon contemplating the probable difficulty of working with a large number of tapered seg-

ments. But a simple jig for your table saw and a novel assembly procedure will enable you to tackle this project—and succeed—with ease.

The planter is designed to hold a 12-inch-high by

MATERIALS LIST		
ITEM	QTY	DESCRIPTION
1	18	$^{11}/_{16}$″ × 2″ × 14″ redwood flat ribs
2	18	$^{11}/_{16}$″ × 1$^3/_4$″ × 14″ redwood tapered ribs
3	18	$^{11}/_{16}$″ × 1$^5/_{16}$″ × 2$^1/_2$″ redwood accent blocks
4	1	$^{11}/_{16}$″ × 9″ × 9″ redwood base

Note: all parts except base are cut from 1 × 3 stock; actual thickness $^{11}/_{16}$″.

ADDITIONAL MATERIALS

Hot-dipped galvanized finishing nails; 17 gauge 1″ brads; #107 rubber bands; white glue (optional); hide glue

ASSEMBLY SEQUENCE

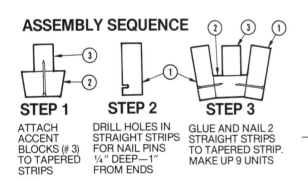

STEP 1
ATTACH ACCENT BLOCKS (# 3) TO TAPERED STRIPS

STEP 2
DRILL HOLES IN STRAIGHT STRIPS FOR NAIL PINS ¼″ DEEP—1″ FROM ENDS

STEP 3
GLUE AND NAIL 2 STRAIGHT STRIPS TO TAPERED STRIP. MAKE UP 9 UNITS

STEP 4
JOIN TWO UNITS TO A TAPERED STRIP TO MAKE UP 4 SUB-ASSEMBLIES AS SHOWN. JOIN THESE TOGETHER WITH A TAPERED STRIP BETWEEN. A SINGLE UNIT AND 2 TAPERED STRIPS GO ON LAST

½″ x ½″ x 1¾″ BLOCK USED WHEN GLUING SECTIONS TOGETHER

1¾″   10°

SPACER BLOCK

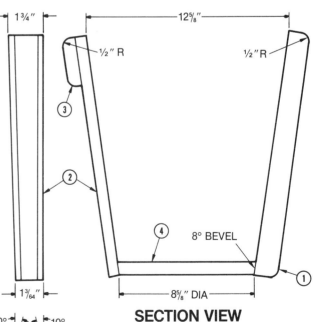

1¾″

12⅝″

½″ R   ½″ R

1³/₆₄″

10°   10°

8° BEVEL

8⅝″ DIA

SECTION VIEW

12-inch-diameter tapered pot with drainage pan. The planter is made of clear, all-heart redwood stock cut into 2-by-14-inch strips. A total of thirty-six strips of this size is required; eighteen have compound angles (beveled tapers), and eighteen are straight. In addition, eighteen blocks ($1^5/_{16}$ by $2^1/_2$ inches) are needed for accenting the top. All these pieces can be cut from 1×3 stock (actual measurement $^{11}/_{16}$ by $2^1/_2$ inches). The bottom panel is cut from a 1×10 board. If you plan to use the planter outdoors it will be necessary to assemble it with waterproof glue.

Novel woodworking techniques are utilized to construct this handsome planter.

The Jig

The jig to cut the beveled tapers is made from a piece of plywood ($3/4 \times 3 \times 25$ inches). To this, tack nail a strip of wood ($1/4 \times 1^1/8 \times 25$ inches) along one edge. Set this strip flush at one corner and allow it to overhang the edge exactly $5/8$ inch at the opposite corner. Mark two lines: one on the strip 6 inches in from the flush corner and a second line 5 inches from the overhanging corner.

Place a square against the angled strip and continue these lines onto the baseboard. Set the table-saw rip fence for a $1^3/4$-inch cut, measured from the inside of the blade to the fence. Make an inside stopped cut in the baseboard, from line to line, to obtain the required off-parallel kerf cut. (Make this cut by elevating the blade into the prepositioned stock.) Remove the guide strip, then saw out the waste so you have a notch 14 inches long to house the redwood strips.

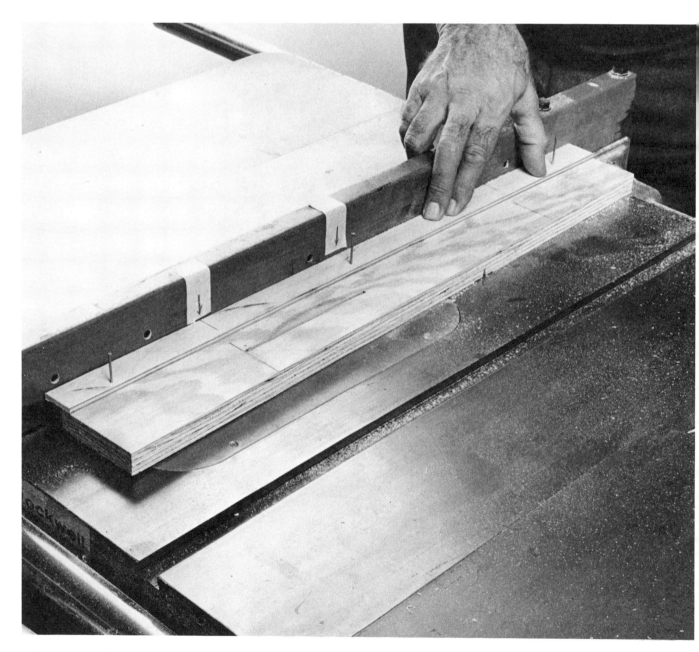

1. The angled guide strip rides against the fence to make the off-parallel partial cut in the jig board. Place the board in position and turn on the power. Then elevate the blade until it cuts through the surface. Advance the piece to the rear mark.

Cutting Compound Angles

Tilt the blade for a 10-degree bevel (tilted toward the fence). Set the fence at $3^1/8$ inch measuring from the blade to the fence at table level. Check the adjustments by cutting a piece of scrap. Insert a workpiece into the jig and make a cut to trim an edge. Make this initial cut on eighteen strips.

Retain one of the waste tapered cutoffs. Flip it over and nail the flat edge to the inside of the jig wall with the thicker end at the narrow end of the jig cutout. Now flip the jig end over end so the narrow portion of the cutout will lead in sawing.

Readjust the fence to $3^1/8$ inches, *minus* the kerf width (this will vary with different blades). Make the second edge cuts on all eighteen beveled strips.

2. Insert a workpiece into the cutout space and make the first pass. Repeat this step with each of the strips. The saw blade is set for a 10-degree bevel.

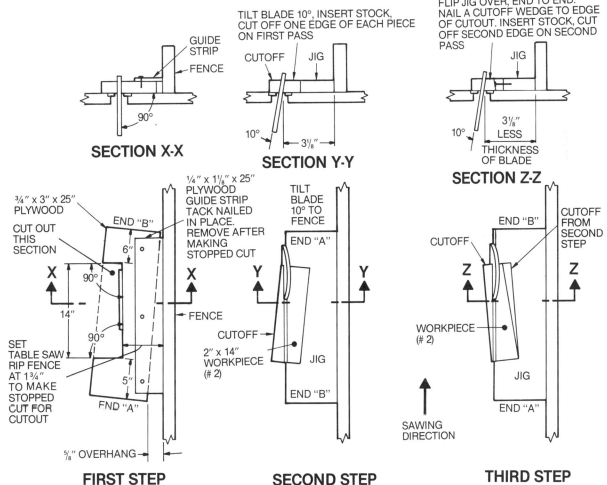

FIRST STEP   SECOND STEP   THIRD STEP

3. Nail one of the waste cutoffs onto the inside edge of the jig cutout before making the second series of cuts. The wide end of the wedge must be located at the narrow part of the cutout to produce the correct angle on the second pass.

4. Make the second pass on each of the eighteen strips to complete cutting the beveled tapers. Insert these strips in the jig so that the first beveled edge mates precisely with the beveled edge in the jig. Masking tape holds the workpiece.

Initial Assembly

Cut the top accent blocks to size, round two ends, and sand. Use two hot-dipped galvanized finishing nails and glue to secure the blocks to the outside top of each tapered strip. Round the two ends of the 14-inch-long divider strips, then prepare for final assembly.

Short, headless nail pins are used to keep all the

5. The first stage of assembly: attach the accent blocks to the tapered strips using glue and two hot-dipped galvanized finishing nails.

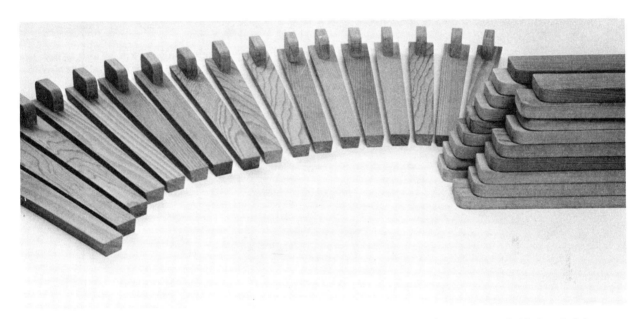

6. All the pieces required prior to the second phase of assembly. In case you are counting, one tapered strip is out of view.

pieces in alignment during assembly. They are inserted near the ends of the straight pieces. To set the pins, bore two pilot holes, 1/4 inch deep, about 1 inch from the ends into the outside surfaces. Do this with a drill press or a drill guide because these holes must be perpendicular to the surface. Use a nail with the head clipped off as a drill bit so the pins will fit snug.

Glue and nail two straight strips to each of nine tapered strips. Clip the heads off 17 gauge 1-inch brads so that you have headless nails about 1/2 inch long. Insert a pin into each pre-drilled hole, blunt end first. Carefully align each of the pre-assembled units to a single tapered strip, then press them together so the nail points will make registration

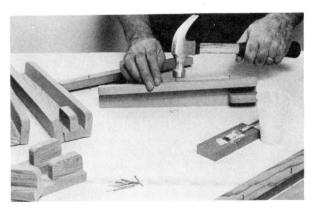

7. Pre-drill the required pilot holes for the nail pins, then proceed to attach two straight pieces to nine of the tapered ones, using glue and nails.

8. Clip off the nail heads. Insert them, blunt end first, into the pre-drilled holes in the faces of the straight strips.

9. Position the tapered strip and press it firmly against the nail points to form registration holes. This is done with the nine un-attached tapered strips.

10. Keep all the point-marked pieces in sequential order and number them accordingly for later rematching.

holes. Besides ensuring alignment, the nail points will prevent the parts from sliding when wetted with glue. Keep all pieces in sequential order and mark the bottoms for later matching.

Temporarily dry-assemble several segments, holding them together with masking tape. Fit the segments together to confirm that assembly will go smoothly.

11. Dry-assemble to test the fit, using masking tape to hold the segments together.

12. This is the important phase of the dry run—the last tapered strip must fit perfectly. If it needs altering you can trim off any excess or build up a slight gap by adding to the edge.

Final Assembly

For this you need thirteen large rubber bands, size # 107, which are $5/8 \times 7$ inches. You can obtain them at most stationery stores. You will also need nine spacing blocks cut from $1/2$-by-$1/2$-inch stock. These blocks, $1^3/4$ inches long, should be angled 10 degrees at each end.

Whereas white glue can be used for the initial pre-assemblies, it should be avoided for the final assembly because it sets too rapidly to allow sufficient working time. A liquid hide glue, which allows ample assembly time, is recommended.

Apply glue to both mating surfaces to make up a segment consisting of two of the partially assembled units with a tapered member between. Use a rubber band stretched lengthwise and a spacer block to hold the butted surfaces in contact. Make up four of these sections.

Now begin the final assembly in the round. Stand the sections on end with the bottom ends up, apply glue, and bring them together. In the final

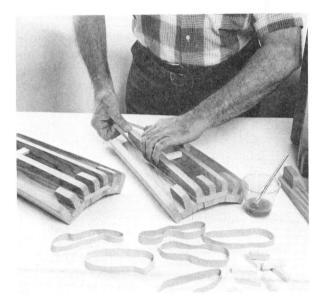

13. Apply glue, locate the nail points into the registration holes, then attach a rubber band lengthwise. Use a spacer block between the center strips to prevent buckling.

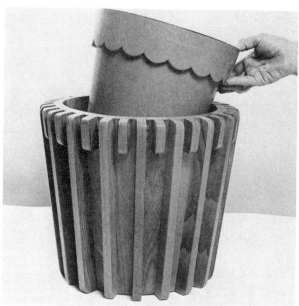

14. Stretch rubber bands around the circumference, remove the spacer blocks, and snip off the lengthwise bands so that all the clamping pressure will be directed toward the center.

15. The planter accepts a 12-by-12-inch pot.

stage you will have one section and two tapered strips to install. Glue in the last section together with a tapered strip. For this it will be necessary to spread apart the assembly slightly to allow the last piece to fit over the projecting nail pins in the adjoining members.

Stretch four of the large rubber bands around the circumference, then cut the vertical rubber bands with scissors so they will not counteract the inward thrust of the circumference bands. Allow the glue to set.

To make a perfectly fitted bottom panel, place the planter onto a piece of paper with a sheet of carbon paper under it, carbon side up. Trace the inside outline, then flip the pattern over and trace the out-

line onto the board. The reason for flipping the pattern is to obtain a true, rather than a reverse, outline. This allows for any minor discrepancy in the shape of the assembly. Set your band, jig, or sabre saw for an 8-degree bevel cut to cut the panel.

If the planter is to be used indoors or in a protected area outdoors, you can apply a clear finish to bring out the rich color of the wood. Clear finishes do not stand up well outdoors and require periodic refinishing. Instead, you can leave the redwood untreated to allow it to weather to an attractive driftwood gray. Or you can apply a water-repellent preservative, which will stabilize the color at a buckskin tan.

20. DUAL-FUNCTION CHESSBOARD

When this attractive board is not in use for play, it hangs on a wall to serve as a decorative accent piece.

The frame is made of crown molding stained red and shaded in black to complement the black-outlined red and natural-wood-toned squares nicely. A novel technique is employed to form and color the squares and outlines neatly so that what

This board is continually in use.

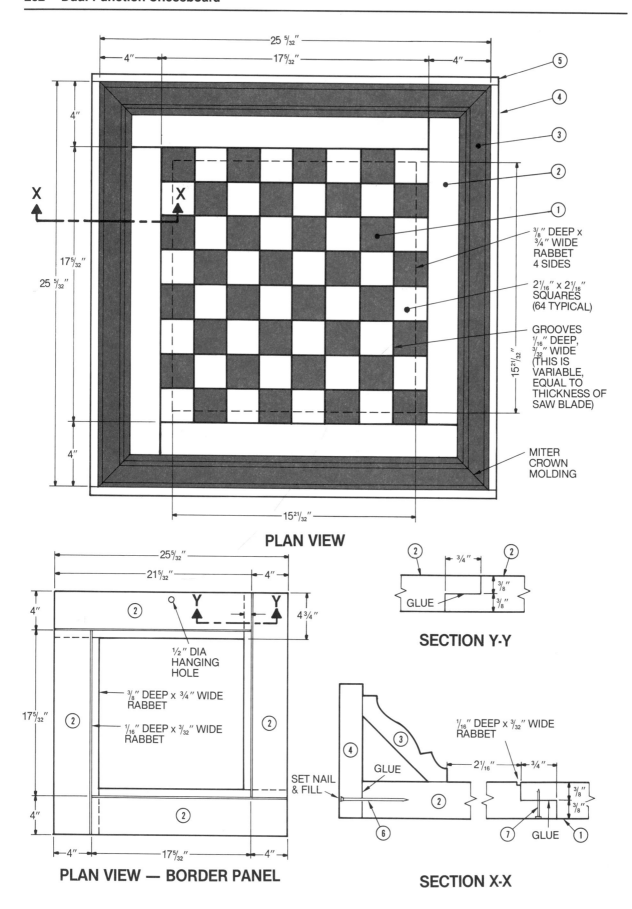

25 5/32"
4"
17 5/32"
4"
5
4
3
2
1

X
X

3/8" DEEP x 3/4" WIDE RABBET 4 SIDES

2 1/16" x 2 1/16" SQUARES (64 TYPICAL)

GROOVES 1/16" DEEP, 3/32" WIDE (THIS IS VARIABLE, EQUAL TO THICKNESS OF SAW BLADE)

17 5/32"
25 5/32"
15 21/32"

4"

MITER CROWN MOLDING

15 21/32"

PLAN VIEW

25 5/32"
21 5/32"
4"
4"
4 3/4"

2
Y
Y

1/2" DIA HANGING HOLE

3/8" DEEP x 3/4" WIDE RABBET

1/16" DEEP x 3/32" WIDE RABBET

17 5/32"

2
2

SET NAIL & FILL

2

4"

4"
17 5/32"
4"

PLAN VIEW — BORDER PANEL

2
3/4"
2

GLUE
3/8"
3/8"

SECTION Y-Y

1/16" DEEP x 3/32" WIDE RABBET

3
4
GLUE
2 1/16"
3/4"
2
6
7
GLUE
1
3/8"
3/8"

SECTION X-X

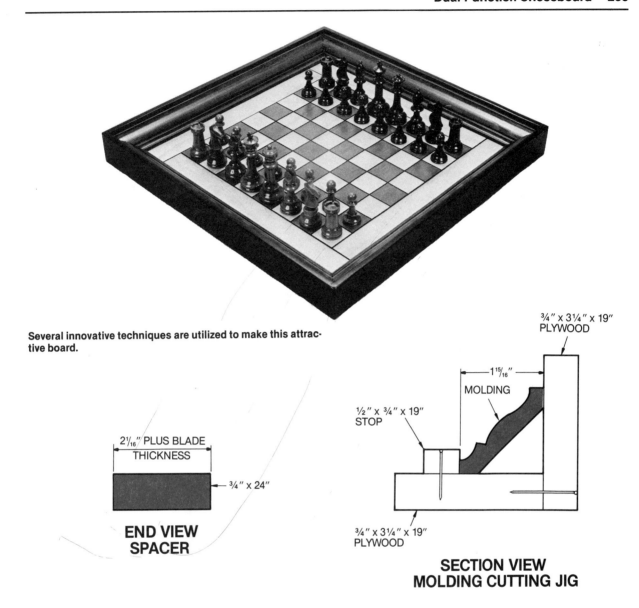

Several innovative techniques are utilized to make this attractive board.

2¹/₁₆″ PLUS BLADE
THICKNESS

¾″ × 24″

**END VIEW
SPACER**

¾″ × 3¼″ × 19″
PLYWOOD

1¹⁵/₁₆″

MOLDING

½″ × ¾″ × 19″
STOP

¾″ × 3¼″ × 19″
PLYWOOD

**SECTION VIEW
MOLDING CUTTING JIG**

DUAL-FUNCTION CHESSBOARD

MATERIALS LIST

ITEM	QTY	DESCRIPTION
1	1	¾″ × 17⁵/₃₂″ × 17⁵/₃₂″ ash veneer plywood
2	4	¾″ × 4¾″ × 21⁵/₃₂″ ash veneer plywood
3	4	3″ × 25⁵/₃₂″ crown molding (actual width approx. 2¹/₂″)
4	2	¹/₂″ × 2³/₄″ × 25⁵/₃₂″ poplar
5	2	¹/₂″ × 2³/₄″ × 26⁵/₃₂″ poplar
6	20	1¹/₂″ finishing nail
7	24	⁵/₈″ nail

ADDITIONAL MATERIALS

White or yellow glue; self-adhesive vinyl sheeting; cranberry stain (squares and molding); black latex paint (molding, grooves, and sides); clear satin finish

would appear to be a most difficult task is really not difficult at all.

The base is made of ³/₄-inch white ash veneer plywood surrounded with a ¹/₂-inch thick solid wood poplar wall. The crown molding may be either poplar or pine, depending on availability.

The squares are made by cutting uniformly spaced, shallow saw kerf grooves. The board and wide border are made up separately in order to allow unobstructed through grooving to form the squares.

Begin by cutting a piece of plywood 18 by 18 inches, perfectly squared. In order to avoid splintering along the edges of the grooves it is essential to use a sharp, smooth-cutting hollow ground blade. On the bottom of this panel clearly mark the letters A and B near two adjoining edges.

Forming Squares

Adjust the rip fence to cut a spacing strip that will have a width of $2^{1}/_{16}$ inches *plus* the thickness of the blade. Put the spacer aside temporarily. To make the grooves, readjust the rip fence for a $2^{1}/_{16}$-inch-wide cut. Adjust the blade height so it projects $^{1}/_{16}$ inch above the table. Make two passes on the face of the panel, first with edge A against the fence, then with edge B. You will now have two grooves at right angles, parallel to the respective edges.

With the last groove cut backed up over the stopped blade, loosen the rip-fence lock and move the fence out of the way. While holding the work in place, butt the spacer strip up to its edge; then slide the rip fence to butt up against the spacer. Lock the fence, lift out the spacer, and make two new kerf cuts opposite edges A and B. Continue making the cuts in this manner, using the spacer to adjust the fence after each pair of cuts. When the seventh sets of grooves have been made, raise the blade, turn the work right side up, and make two final cuts to trim off the waste. The board will now have sixty-four precisely dimensioned squares.

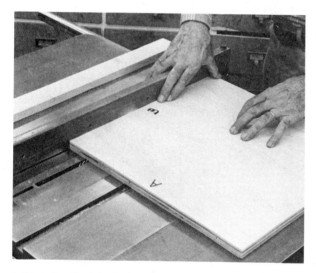

1. Blade is raised $^{1}/_{16}$ inch to make the shallow kerf cuts. One pass is made along sides A and B. The spacer strip is not used to set up these two initial cuts.

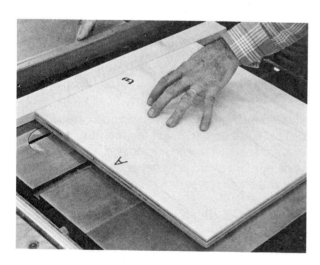

2. Spacer strip is used to reset the rip fence for the subsequent pairs of cuts. The panel is held in place over the blade while the fence is positioned.

3. Spacer is withdrawn and the panel is backed off the blade and moved against the fence to make the new cut.

4. This is how the grooves progress.

5. Blade is raised for the two final cuts to drop off waste.

Border Panels

Cut the four plywood border panels next. Note: a shallow rabbet sized to match the grooves in the board is required on the top inside edge of each of the panels. These should be cut with the same smooth-cutting blade as was used for the grooving operation. Set the rip fence for a cut to a width of ³/₄ inch *plus* the thickness of the blade. After these cuts have been made, switch over to a dado blade and cut the rabbets on the faces of the border pieces and on the back of the baseboard. When the panels are assembled the shallow rabbets on the border panels will automatically become grooves.

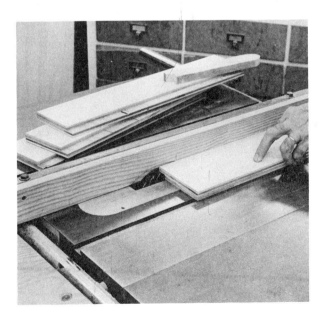

6. Auxiliary fence board is used to make inboard rabbet cuts with a dado head.

7. The miter gauge is used to guide the work for the end rabbets. A long face board on the miter gauge helps to hold the pieces true.

Now glue the borders to the base. Turn all the pieces bottom side up and work on a true, flat surface. Start two ⁵/₈-inch nails into the overlaps to form registration holes. Apply glue and nail the parts together. Go easy with the glue along the shoulders of the lap joints to avoid squeeze-out onto the faces of the panels and into the grooves. Sand the face of the assembled panel, then attach three of the side wall members. The fourth is installed later.

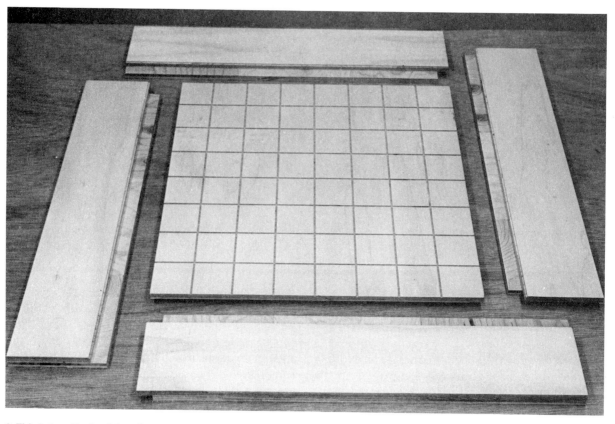

8. This is how the lap joints line up in assembly position.

9. Joints are held with glue and ⁵/₈-inch nails. The nails serve for assembly only, not strength. Apply glue sparingly near the finish surfaces to avoid marring the work.

Decorating

Masking for controlled staining of the squares is done with self-adhesive vinyl sheeting. Cut four pieces to mask the border areas, then cut thirty-two squares 2$\frac{1}{8}$ by 2$\frac{1}{8}$ inches. Peel off the backing and apply the masks to each alternate square on the board. Press firmly along the edges to make good contact so the stain will not seep into where it is not wanted.

Use transparent Deft cranberry stain to color the squares. Work only half the board at a time because this stain sets rather quickly. This stain is water-based, so wipe off the excess with a damp cloth. Allow it to dry thoroughly before peeling off the masking. Follow with three full coats of Deft clear satin finish; sand lightly between coats.

Use black latex paint to color the grooves. Since the heavy finish will prevent penetration of the paint into the surface you need not be too fussy about accuracy in applying the paint in the grooves. But you must work quickly.

Use a brush with short bristles and work the paint into the grooves. Do only two or three rows at a time because if the quick-drying latex paint sets on the surface it will be difficult, if at all possible, to remove. Immediately after the grooves have been coated, use a damp, soft cloth wrapped around a block of wood to clean the excess paint from the surface. The paint will remain in the grooves resulting in perfectly sharp outlines.

10. Self-adhesive vinyl sheeting is used for masking the wood prior to selective staining of the squares.

11. Alternate squares are covered, then burnished with the finger to obtain good contact. The border panels are also masked as well.

12. A coating of the red stain is laid on; then the residue is wiped off with a water-dampened cloth. Work no more than half the board at a time.

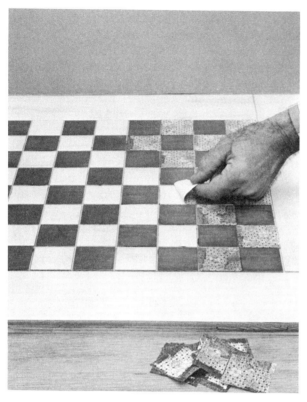

13. Vinyl squares are peeled off by pulling at a low angle to avoid lifting any veneer.

14. Three coats of clear finish are applied to protect the surface from paint penetration.

15. A short-bristled brush is used to work the black paint into the grooves. Here again, work only part of the board at a time.

16. A pad of damp cloth wrapped over a block of wood is used to wipe the paint from the surface.

The Frame

Crown molding requires a compound angle cut to form a mitered corner joint. Unless you use a simple, special jig you may have difficulty with this cut. The Crown Molding Miter Box on page 3 works well for this purpose, but you may want to make the one shown here, which will serve the immediate need. It holds the molding tilted at the installation angle. When a 45-degree miter cut is made with the molding in this position the result will be the required compound miter. The saw blade is not tilted for these cuts.

Mark the molding lengths on the top edges and align the marks with the edges of the jig to make accurate cutoffs. The molding is assembled into a frame and stained prior to installation. To assemble the frame, size the absorbent end-grain mitered

17. Jig for mitering the crown molding must hold it in its final installation attitude. A nail point projecting slightly from the base keeps the workpiece from sliding during cutting.

faces by applying a light coat of white or yellow glue. Allow it to set for about 10 minutes. Lay down scraps of waxed paper at the corners of the board to protect against glue run-off. Apply glue to the ends of the molding strips and set them in place on the board; use masking tape to hold them in position. When the glue has set, remove the frame from the board and apply the finish. Now you can install the fourth side wall. This piece had been left off to this point to allow easier handling of the molding strips during checking and gluing. Apply black paint to the top and inside upper edges of the wall and to the outside faces.

To color the frame, apply the red stain and allow it to dry. Then, working one section at a time, apply black latex paint and follow by wiping the black from the high spots with a damp cloth padded over a block of wood. This will result in black shading in the depressed areas of the molding.

Insert the frame, then apply three coats of clear satin finish over the entire chessboard. Glue won't be needed to secure the frame in place if it is well fitted because the finish coats will bond it.

18. Miter gauges sometimes go out of alignment. A triangle is used to adjust the gauge to make the 45-degree angle cuts on the jig ends.

19. Cuts are made on one end of each piece. The miter gauge is then shifted to the opposite angle.

20. Frame is assembled in place, but it is not permanently attached at this time. Tape is used to hold the parts while the glue sets. Waxed paper under the corners blocks glue run-off.

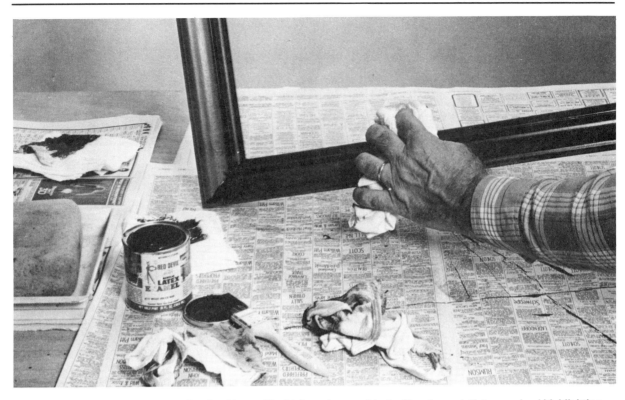

21. Black paint is applied over the red-stained frame. The high spots are rubbed with a damp cloth to reveal red highlighting and black shading.

22. Frame is set into place after the top edges of the side walls have been painted. This results in a clean-cut demarcation between the red frame and black border.

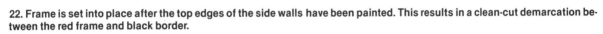

21. GOLD-TOOLED DESK TOP

For a fine touch of charm and beauty you might consider incorporating a gold-tooled simulated leather top on your next desk or occasional-table project. Using the materials and procedures shown will result in a look quite like that of real leather—so much so that perhaps only an expert could tell the difference.

The grained "leather" shown is Naugahyde. It is a vinyl fabric with a remarkable resemblance to fine leather in appearance and feel. It is commonly available at fabric shops in 54-inch widths in a variety of colors and textures. There are several grades of the material, and it is important to obtain the Service Quality grade because that has the proper thickness for this application.

In the following directions the Naugahyde will usually simply be referred to as leather.

In designing the top for a project, plan on utilizing a frame construction with deep rabbets into which the top panel with the leather pre-mounted can be set in flush. This is a simplified and very practical method for inlaying the material. The advantage is that the leather can be cemented easily and edge-trimmed to perfection on the unattached top panel prior to installation. Also, the border decorating can be done before final assembly. If an error is made during this phase, all will not be lost, you'll need only to make a new insert panel. And further, the surrounding frame can be sanded and finish-coated without danger of marring the inlay.

The drawing shows a typical cross section for a framed construction top with a plywood panel insert. When rabbeting the recess to receive the panel, allow $1/16$ inch extra depth for the thickness of the leather.

To make the inlay panel, rough trim the leather so it overhangs the plywood an inch or so. Apply contact cement to both surfaces and allow to dry. Naugahyde has a knitted fabric backing that somewhat slows the drying of the cement, so it is advisable to coat it about $1/2$ hour before coating the wood. In this way both surfaces will be properly set for assembly at about the same time. Use two sheets of wrapping paper as a separator while positioning the material on the wood. Slip out the first piece to allow partial contact, then remove the other. Use a small 3- or 4-inch rubber roller to make positive contact throughout.

Turn the panel over and use a sharp knife to trim off the waste flush with the edge of the plywood.

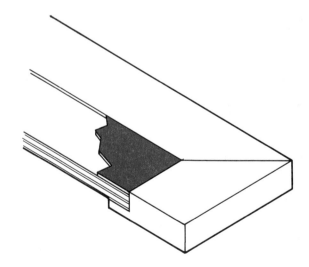

A desk or table top made to receive the inlay panel should be rabbeted or otherwise recessed $1/16$ inch deeper than the thickness of the backboard to allow for the thickness of the leather. The leather top can be flush or set slightly below the surrounding surface, but it should not be higher.

Gold tooling is a specialized art of the leatherworker, but you can readily learn how to duplicate a piece such as this by following the simplified technique shown here.

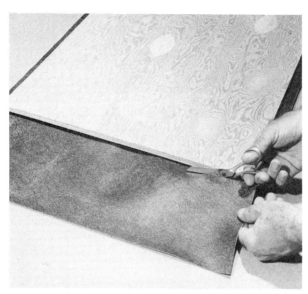

1. Trim the "leather" about an inch oversize and save a fair-sized piece of scrap for making test impressions.

2. Use contact cement to bond the leather to the wood. Apply it to both surfaces and allow it to set until it is no longer tacky. The nonflammable type should be used in the home or shop.

3. Use paper slip sheets to align the leather to the wood; it must overhang slightly all around. Let the material down gradually after it has made contact at one end. Sweep across the surface with the hand to prevent air pockets.

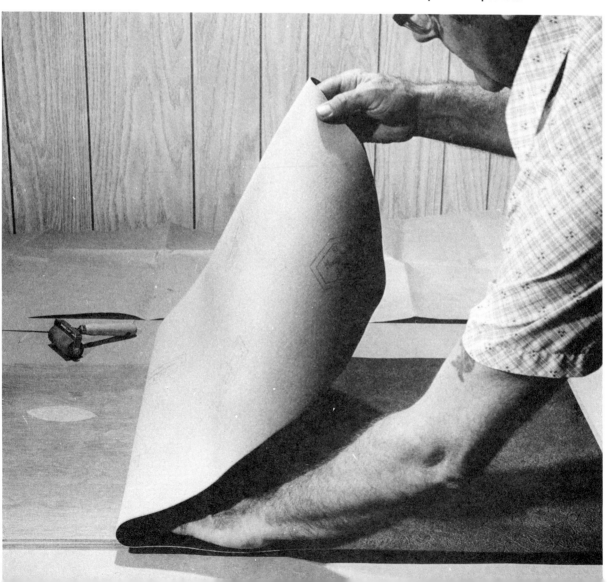

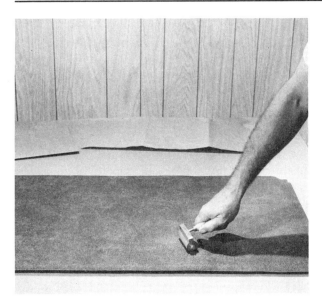

4. Use a small rubber roller to make good contact over the entire surface.

5. Turn the piece over and trim along the edge of the board with a very sharp knife to remove the excess.

Cement a leftover piece of leather to a scrap board for use during the next phase.

To do the gold tooling, you will need an alcohol lamp, a decorative leathercraft stamp, a roll of 1-inch-wide mylar gold foil, and some masking tape. Use only alcohol to fuel the lamp; kerosene and other fuels leave a heavy soot that would transfer from the tool to the work. A candle flame is not suitable for the same reason. The stamp shown, # V 417 is available at hobby or leathercraft shops. Other designs are also available. The mylar gold foil is a trade specialty item that may not be readily available at local sources, but you can order it by mail from the source listed at the end of these directions. In order to avoid burning your fingers, you should make a handle for the stamping tool by boring a hole in the end of a small piece of ³/₄-inch-square stock.

6. You need an alcohol lamp, a stamping tool, and gold foil to make the decorative border. A homemade wood handle protects fingers from the hot tool.

Tooling

Start by cutting and taping strips of the foil into place, dull side down. Do *not* overlap the corners; they must butt evenly. Apply the tape close to the edges so it will not be in the way of the design. The stamp design is crescent-shaped, so a double impression is made to obtain the completed pattern shown. The first impression, which forms one half of the pattern, is made against a strip of wood to ensure obtaining a straight line of scallops. Tape the guide strip in place so it covers exactly half the foil, lengthwise.

The tool makes an impression 1 inch long. In order to end up with a full loop at the ends of a run it is advisable to make pencil marks, 1 inch apart, on the guide strip. Center the guide accordingly. If the inside cross borders do not divide evenly into full inches you can obtain a partial impression by inserting a few sheets of paper between the foil and the stamp.

Heat the stamp over the flame for about 10 seconds. Set up the scrap piece of leather with a strip of foil in place and make an impression by pressing the stamp squarely into it. If the tool is too hot it will scorch the foil and the leather as well. Quickly make another test impression, lift the foil and check the result. If it looks good, proceed to make a series of impressions on the work. Position the tool so both points of the crescent butt up against the guide. Press down firmly and hold the tool in place for about one second for each impression. Each heating of the tool will be sufficient for about five or six impressions. Do not go beyond this, because a cool tool will not transfer the gold. Be sure to make a test impression each time the tool is reheated.

When the initial row of scallops is completed, remove the guide strip and continue the impressions freehand by matching up the crescent points. When all the stamping has been completed, peel away the foil to reveal a permanent gold-impressed pattern.

Note: the gold mylar foil is available from Armor Products, PO Box 290, Deer Park, NY 11729. If you have difficulty in obtaining the leathercraft stamp, write to this address: Tandy Leather Company, 3 Tandy Center, Fort Worth, TX 76101, for source information.

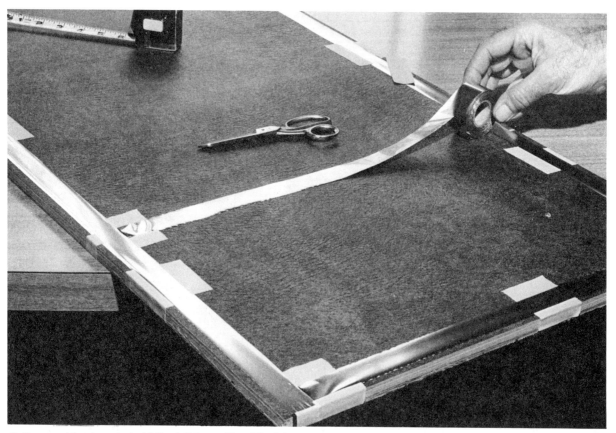

7. Use masking tape to hold the foil in position. The joints must be butted, not overlapped.

8. A wood strip guides the tool in a straight line for the first row of scallops. Note: the scrap piece behind the lamp is used after each heating in order to check the working temperature of the stamp.

9. The second row of impressions is made freehand after removing the guide strip.

10. Peel off the foil to reveal the completed border design.

22. OAK HALL BUTLER

Make this impressive piece of furniture to serve as a functional focal point in your foyer. The project will afford a good opportunity to practice some relatively simple basic joinery techniques.

Featured on the recessed-panel backboard are a fair-sized mirror, a colorful opalescent stained-glass panel, and Victorian-styled brass and porcelain clothing hooks. The chest section includes a roomy drawer and a storage compartment under the seat for foul-weather footwear. An aluminum pan lines the bottom of this compartment to contain any water run-off from wet shoes.

The stock required for the backboard and chest is 4/4 solid oak and 1/4-inch oak veneer plywood. Two pieces of 5/4 oak are used to build up the cornice. Other stock requirements are 3/4-inch fir plywood (for the chest back and bottom), a piece of 2 × 2 softwood (to back up the skirt), and 1/2-inch poplar and plywood (for the drawer). The usual thickness of 4/4 stock is 13/16 inch, thus the plan is

sized accordingly. If the actual thickness of your stock differs you will have to alter the dimensions to suit your own requirements.

Aluminum-lined chest stores rainy weather footwear.

Detail view of the opalescent stained-glass accent panel.

Several basic woodworking skills and techniques are utilized to produce this charming piece.

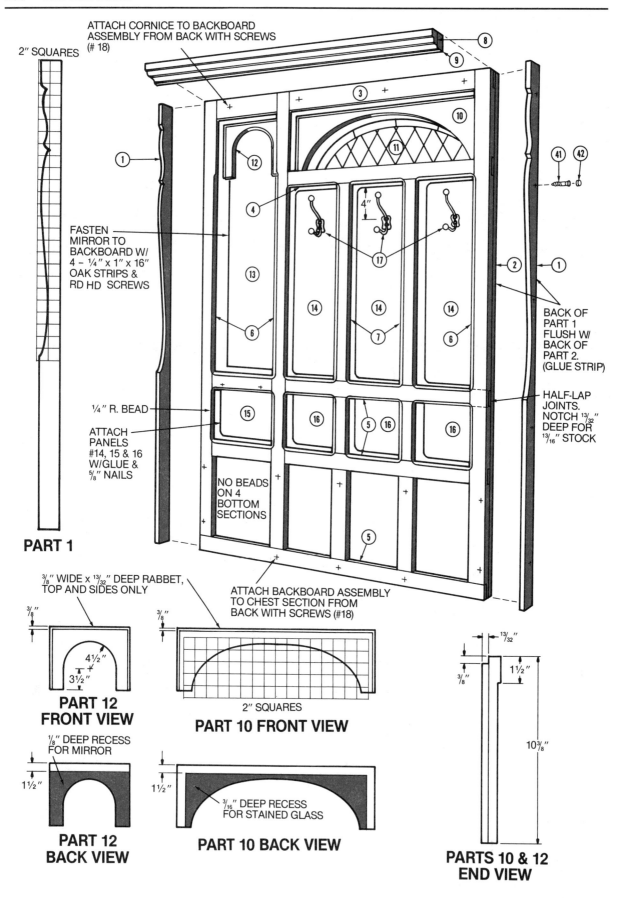

2″ SQUARES

ATTACH CORNICE TO BACKBOARD
ASSEMBLY FROM BACK WITH SCREWS
(# 18)

FASTEN
MIRROR TO
BACKBOARD W/
4 – ¼″ x 1″ x 16″
OAK STRIPS &
RD HD SCREWS

¼″ R. BEAD

ATTACH
PANELS
#14, 15 & 16
W/GLUE &
⅝″ NAILS

NO BEADS
ON 4
BOTTOM
SECTIONS

PART 1

BACK OF
PART 1
FLUSH W/
BACK OF
PART 2.
(GLUE STRIP)

HALF-LAP
JOINTS.
NOTCH ¹³/₃₂″
DEEP FOR
¹³/₁₆″ STOCK

ATTACH BACKBOARD ASSEMBLY
TO CHEST SECTION FROM
BACK WITH SCREWS (#18)

⅜″ WIDE x ¹³/₃₂″ DEEP RABBET,
TOP AND SIDES ONLY

⅜″

4½″
3½″

**PART 12
FRONT VIEW**

⅛″ DEEP RECESS
FOR MIRROR

1½″

**PART 12
BACK VIEW**

⅜″

2″ SQUARES

PART 10 FRONT VIEW

1½″

³/₁₆″ DEEP RECESS
FOR STAINED GLASS

PART 10 BACK VIEW

¹³/₃₂″

1½″

⅜″

10³/₈″

**PARTS 10 & 12
END VIEW**

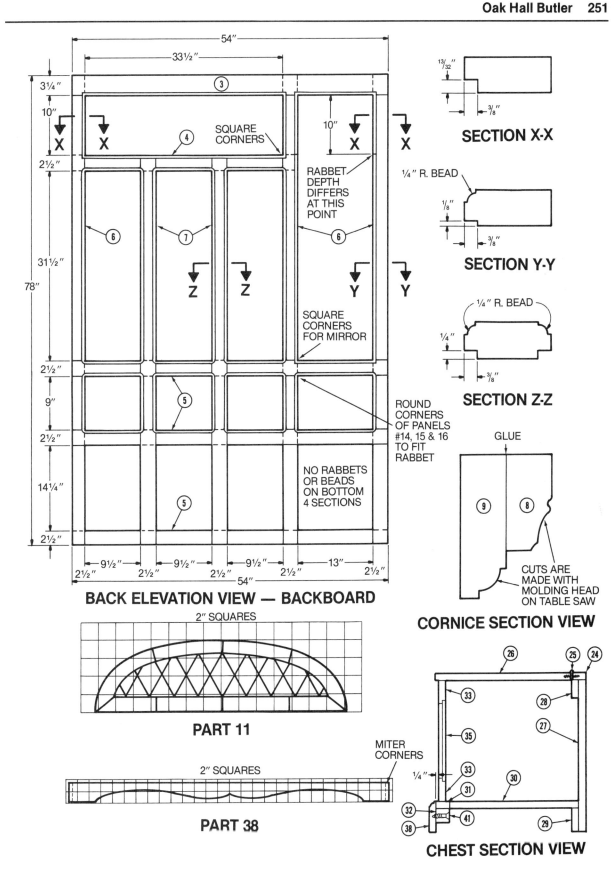

SECTION X-X

¹³/₃₂″

³/₈″

¼″ R. BEAD

⅛″

³/₈″

SECTION Y-Y

¼″ R. BEAD

¼″

³/₈″

SECTION Z-Z

GLUE

9 8

CUTS ARE MADE WITH MOLDING HEAD ON TABLE SAW

CORNICE SECTION VIEW

54″

33½″

3

3¼″

10″

4

X X X X

SQUARE CORNERS

RABBET DEPTH DIFFERS AT THIS POINT

2½″

6 7 6

31½″

78″

Z Z Y Y

SQUARE CORNERS FOR MIRROR

2½″

9″

5

ROUND CORNERS OF PANELS #14, 15 & 16 TO FIT RABBET

2½″

14¼″

5

NO RABBETS OR BEADS ON BOTTOM 4 SECTIONS

2½″

9½″ 9½″ 9½″ 13″

2½″ 2½″ 2½″ 2½″ 2½″

54″

BACK ELEVATION VIEW — BACKBOARD

2″ SQUARES

PART 11

2″ SQUARES

MITER CORNERS

PART 38

26 25 24

33 28

27

35

¼″ 33 30

31

32 41

38 29

CHEST SECTION VIEW

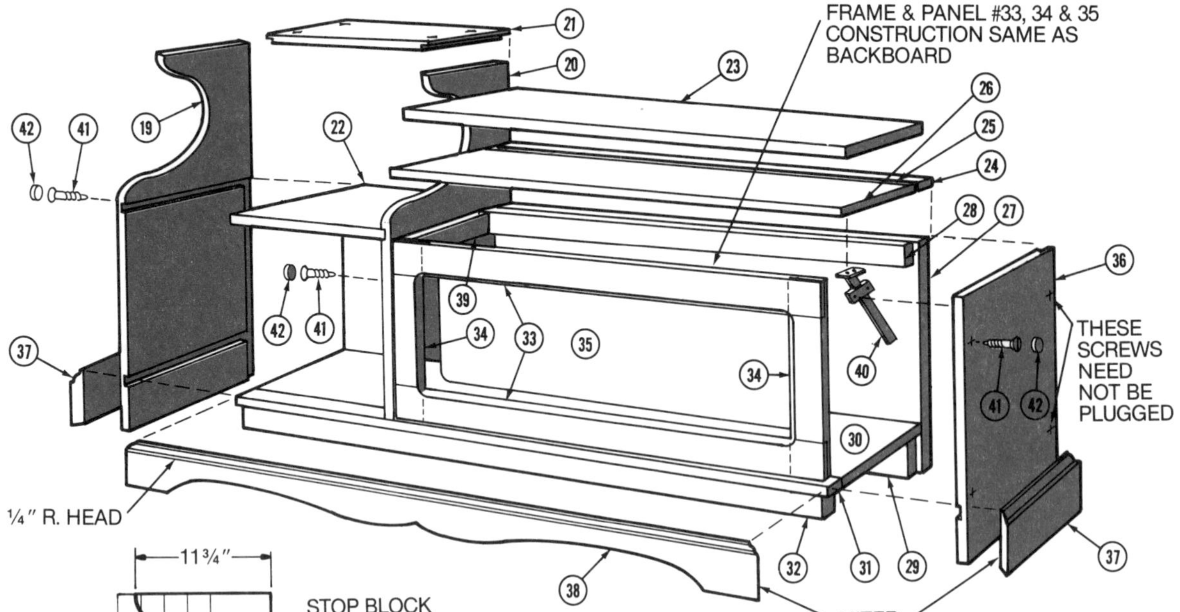
FRAME & PANEL #33, 34 & 35
CONSTRUCTION SAME AS
BACKBOARD
THESE
SCREWS
NEED
NOT BE
PLUGGED
¼″ R. HEAD
MITER
CORNERS
21
20
23
26
25
24
28
27
36
42
41
19
22
39
42
41
34
33
35
40
34
41
42
37
30
32
31
29
37
38

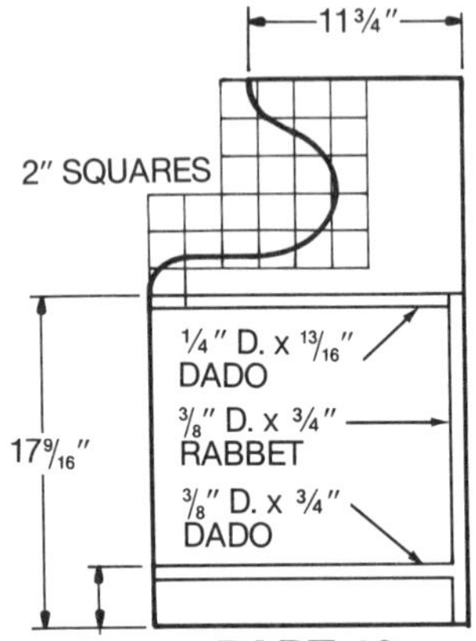
11¾″
2″ SQUARES
¼″ D. x 13/16″
DADO
⅜″ D. x ¾″
RABBET
17 9/16″
⅜″ D. x ¾″
DADO
3⅜″

PART 19

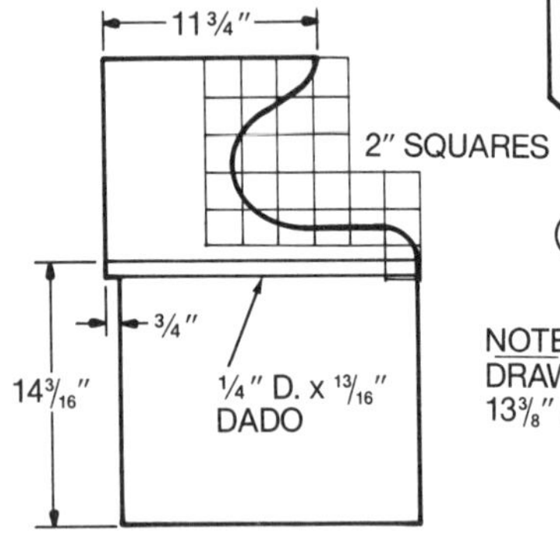
11¾″
2″ SQUARES
¾″
14 3/16″
¼″ D. x 13/16″
DADO

PART 20

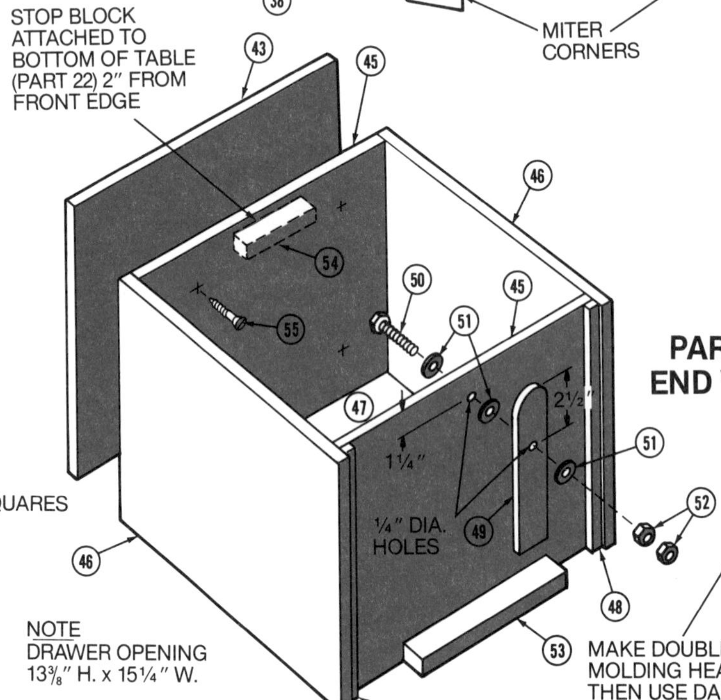
STOP BLOCK
ATTACHED TO
BOTTOM OF TABLE
(PART 22) 2″ FROM
FRONT EDGE
43
45
46
54
50
45
55
51
47
2½″
1¼″
51
¼″ DIA.
HOLES
49
52
46
48
53
48
NOTE
DRAWER OPENING
13⅜″ H. x 15¼″ W.

DRAWER

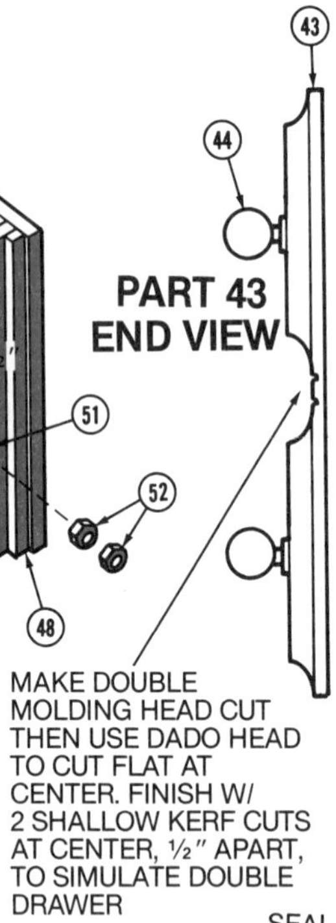
43
44
PART 43
END VIEW
MAKE DOUBLE
MOLDING HEAD CUT
THEN USE DADO HEAD
TO CUT FLAT AT
CENTER. FINISH W/
2 SHALLOW KERF CUTS
AT CENTER, ½″ APART,
TO SIMULATE DOUBLE
DRAWER

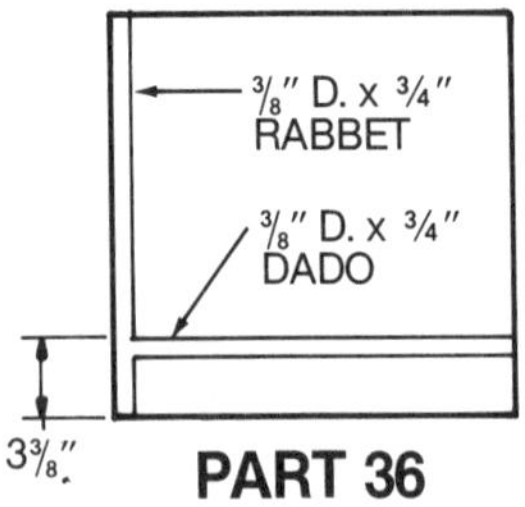
⅜″ D. x ¾″
RABBET
⅜″ D. x ¾″
DADO
3⅜″

PART 36

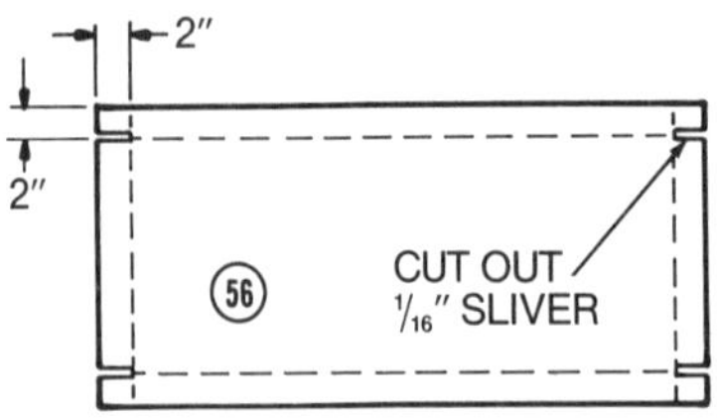
2″
2″
56
CUT OUT
1/16″ SLIVER

RAIN SHOE PAN

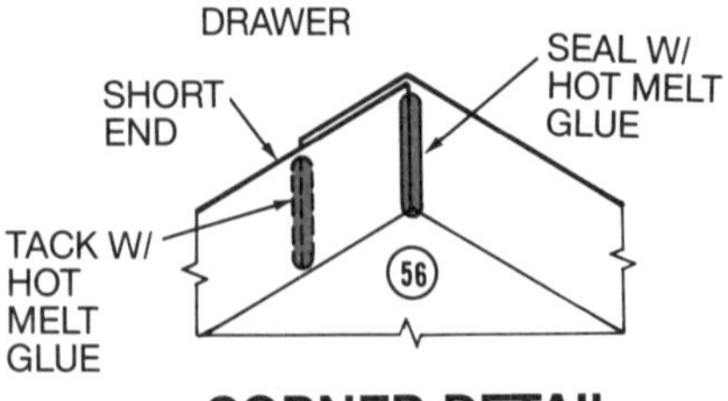
SHORT
END
SEAL W/
HOT MELT
GLUE
TACK W/
HOT
MELT
GLUE
56

CORNER DETAIL

Back Panel Construction

The back panel is assembled with end and middle half-lap joints. Note: the joint lines are *not* the same throughout; they are variously vertical and horizontal in order to achieve balanced flow lines of the rails and stiles—horizontal and vertical members, respectively.

Rip the frame stock to width, then cut the pieces to the required finished lengths. Lay the strips in their respective positions on the floor, measuring carefully so they will be accurately spaced. Check the corners for true right angles with a large square. Use a sharp pencil or knife to mark the intersections for the lap joints. The marks can be made completely across the upper surfaces, but you will not have complete access across the lower faces so tick

marks on the edges will have to do. The lines can be completed when the layout is disassembled. Be sure to identify each mating piece in advance.

Use a dado head on a saw with the blade height adjusted to cut through exactly one half the thickness of the stock. When all the pieces have been notched, make a dry-assembly to check that they fit together properly before gluing.

The glue assembly is best accomplished on a flat plywood panel set on sawhorses. Lay down pieces of waxed paper under all joints so they will not become glued to the work surface. Start out by laying down the two stiles that abut the stained-glass section. Insert the second and third rails (counting from the bottom) as shown in the photo. Next, lay

OAK HALL BUTLER
MATERIALS LIST

ITEM	QTY	DESCRIPTION
1	2	$13/16'' \times 31/2'' \times 78''$ oak
2	2	$1/2'' \times 1/2'' \times 78''$ oak
3	1	$13/16'' \times 31/4'' \times 54''$ oak
4	1	$13/16'' \times 21/2'' \times 381/2''$ oak
5	3	$13/16'' \times 21/2'' \times 54''$ oak
6	3	$13/16'' \times 21/2'' \times 78''$ oak
7	2	$13/16'' \times 21/2'' \times 643/4''$ oak
8	1	$11/6'' \times 21/4'' \times 54''$ oak
9	1	$11/6'' \times 35/16'' \times 54''$ oak
10	1	$13/16'' \times 103/8'' \times 431/4''$ oak
11	1	stained glass panel (See Text)
12	1	$13/16'' \times 103/8'' \times 133/4''$ oak
13	1	$1/8'' \times 135/8'' \times 431/8''$ mirror
14	3	$1/4'' \times 101/4'' \times 321/4''$ oak veneer plywood
15	1	$1/4'' \times 93/4'' \times 133/4''$ oak veneer plywood
16	3	$1/4'' \times 93/4'' \times 101/4''$ oak veneer plywood
17	3	barber shop hooks
18	14	2" # 12 fh wood screws
19	1	$13/16'' \times 171/4'' \times 283/4''$ oak
20	1	$13/16'' \times 171/4'' \times 253/8''$ oak
21	1	$13/16'' \times 111/4'' \times 167/8''$ oak
22	1	$13/16'' \times 153/4'' \times 171/4''$ oak
23	1	$1'' \times 161/2'' \times 361/2''$ foam with $35'' \times 38''$ cloth cover
24	1	$13/16'' \times 15/8'' \times 371/8''$ oak
25	1	$11/2'' \times 36''$ strip hinge
26	1	$13/16'' \times 151/2'' \times 367/8''$ oak
27	1	$3/4'' \times 163/4'' \times 531/4''$ fir plywood
28	1	$13/16'' \times 2'' \times 36''$ oak

MATERIALS LIST

ITEM	QTY	DESCRIPTION
29	1	$3/4'' \times 25/8'' \times 521/2''$ fir plywood
30	1	$3/4'' \times 15'' \times 531/4''$ fir plywood
31	1	$3/4'' \times 11/2'' \times 531/4''$ oak
32	1	$11/2'' \times 11/2'' \times 521/2''$ fir plywood
33	2	$13/16'' \times 21/2'' \times 361/4''$ oak
34	2	$13/16'' \times 21/2'' \times 133/8''$ oak
35	1	$1/4'' \times 91/8'' \times 32''$ oak veneer plywood
36	1	$13/16'' \times 163/4'' \times 171/4''$ oak
37	2	$13/16'' \times 31/4'' \times 16''$ oak
38	1	$13/16'' \times 31/4'' \times 551/2''$ oak
39	1	$13/16'' \times 2'' \times 141/2''$ oak
40	1	friction type lid stay
41	26	2" # 10 fh wood screws
42	22	oak plugs to fit screw holes
43	1	$13/16'' \times 131/4'' \times 151/8''$ oak
44	2	porcelain knob
45	2	$1/2'' \times 12'' \times 141/8''$ poplar
46	2	$1/2'' \times 12'' \times 15''$ poplar
47	1	$1/2'' \times 131/4'' \times 141/8''$ fir plywood
48	2	$1/2'' \times 1/2'' \times 12''$ poplar
49	1	$1/2'' \times 15/8'' \times 61/2''$ poplar
50	1	$1/4'' \times 11/2''$ hex-hd bolt
51	3	$1/4''$ flat washer
52	2	$1/4''$ hex nut
53	1	$3/4'' \times 11/4'' \times 8''$ wood block
54	1	$11/8'' \times 11/8'' \times 4''$ wood block
55	4	1" # 8 fh wood screw
56	1	$19'' \times 391/2''$ light gauge aluminum sheet (.020)

Note: $13/16''$ is the common actual thickness of 4/4 stock.

1. Frame members are cut to size and laid out in precise position to mark the half-lap joints. Be sure to identify each member to avoid later mix-ups.

2. Completed half-lap notches in the recessed-panel frame members. Notice how the notch positions vary.

3. To avoid hang-ups the assembly should begin with these four pieces. This view is true-oriented—the lower right of the frame is in the lower right of the photograph.

4. Bowed board applies pressure to hard-to-reach inner joints. The center spacer is thicker than the others.

5. Router with a 3/8-inch rabbeting cutter is used to cut the recesses for the plywood panel inserts.

6. A mortising bit is used to cut the recess for the stained-glass panel. Router is worked from the outside in to preserve the flat support for the router base. Nailed-wood stops control the lateral movement of the router.

in the short rail that forms the bottom of the glass-panel frame. The top rail follows, then the three remaining stiles. The bottom rail goes on last. This sequence will avoid any hang-ups.

In order to obtain clamp pressure in the central area where clamps will not reach, place wood strips over the center joints with a double thickness over the middle rail. Bridge these strips with a long board, then clamp the board down at both ends. The bowed, tensioned board will impart pressure at the joints under the strips. It is important to work on a level, flat surface in this manner; any twist that develops in the frame may remain permanently after the glue has set.

The assembled frame is rabbeted on the back for the plywood panel inserts and for the solid wood arches and mirror. The router, with a 3/8-inch rabbeting bit, is used for this. This results in round inside corners. The corners are left round in the sections that will receive the plywood panels—the panels are simply cut with rounded corners to fit. The corners in the sections that receive the solid wood inserts and the mirror must be squared with a chisel. Notice in the drawing that the rabbet in the vertical mirror section is cut to two depths in order to accommodate the arched panel and the mirror. The arched panel for the mirror can be recessed with a dado head on the saw, but the one for the stained glass must be recessed with a router because three sides must remain at full thickness.

Glue in the arched panels, then turn the frame right side up to permit cutting the molded edge around all the openings, except the four at the bottom. Use the router with a 1/4-inch bead bit for this. Note: it is very important to move the router in the

7. Detail of the arched insert panels for the stained glass and mirror. Notice the two rabbet levels in the mirror frame section.

8. Bead molded edge is cut after the arched insert panels have been glued in place. Edges in the four bottom sections are not shaped since they are concealed by the chest.

9. A flexible sanding wheel is used to ease the sharp corners of the molded edges. Take care not to overdo.

10. Plywood insert panels are cut with rounded corners to match the round-cornered rabbets. This avoids much hand-chiseling. Drill pilot holes for the nails and use just enough glue to wet the surfaces. If excess glue runs onto the front it will be difficult to remove.

11. The band saw is best to cut the contoured outlines. This can also be done with a scroll or sabre saw, if necessary.

opposite direction of the rotation of the cutter in order to avoid splintering the stock ahead of the cut.

Attach the plywood inserts with glue and 5/8-inch nails after all the parts have been sanded. Pre-drill pilot holes for the nails using one of the nails with the head clipped off as a drill bit. Bore these holes 3/8 inch deep. Apply the glue sparingly to avoid run-off onto the face of the work.

Make the cornice by joining two pieces of 5/4 stock that are individually pre-shaped with a molding head cutter on a table saw. Several passes over two different cutters with fence positions shifted between the cuts results in contours of greater variety.

Secure the backboard side strips with glue and screws that are recessed in counterbored holes. Cut wood plugs from the same stock and use them to conceal the screws. Note: the edges of the side pieces project 1/2 inch beyond the back. This is to allow the unit to snug up to a wall even if a bulge is present in the wall.

12. Spokeshave is used to smooth out the saw ripples and any irregularities. Change the stroke direction when necessary so the cut is always made with the grain.

The Chest Section

Solid lumber stock in the widths required for the chest members will have to be made up by edge-gluing two or more boards. Use dowels or splines to reinforce the butt joints. (Fir plywood is less costly than hardwood and is used for the back and bottom in the interest of economy.) A facing strip of solid oak is added to the front edge of the bottom panel because a small portion of it will be exposed to view. Cut the parts for the chest to size, but do not cut the contours until the dadoes and rabbets have been made.

Assembly begins with the back and bottom panels. Use 2-inch finishing nails to attach the back to the bottom. Glue in the cleat as indicated to reinforce this joint. Dry-assemble the oak members and hold them together with clamps while boring screw pilot holes. Attach the left-hand member first. Follow this with the drawer compartment top and the right end of the drawer compartment, then add the top shelf.

Make up the front panel frame and clamp it in place as shown to bore the screw pilot holes. Do this *before* shaping the inside edges. The skirt members are joined with 45-degree miters. Cut the miters *before* shaping the edges or cutting the contours in the front piece.

The drawer front is made to simulate a double drawer by means of molded cuts. A molding head is used on the saw to make the required border and center-division cuts. When the shaped cuts have been made, two parallel thin-blade kerf cuts are made across the center to form a simulated drawer divider. The drawer is constructed of 1/2-inch poplar as a four-sided box around a 1/2-inch plywood bottom. The front is then attached to it.

13. Dado head is used to cut the dadoes and the stopped rabbets. The auxiliary wood fence permits inboard rabbeting against the fence. Curved end of the stopped cut is squared with a chisel.

14. Chest members ready for assembly. All the inside surfaces should be finish-sanded *before* assembly.

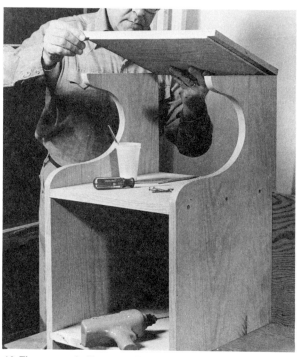

15. Order of assembly is important. The bottom is attached to the back, then the end panel is added. The drawer top goes on next, then the right drawer panel.

16. The upper shelf goes on last. All the screw holes are counterbored and plugged.

17. Frame for the front recessed panel is clamped in place and drilled for the screw holes. This is done before the edges are molded to permit solid clamping.

18. Cutting the 45-degree miter on the skirt piece. The saw is kept in the same bevel position and the work is flipped end to end to obtain the opposite angle cut.

19. Two cuts are made with a molding cutter head to simulate the double drawer. This is a test piece only. When making this panel be sure to make the end-grain cuts first to avoid splintering at the corners.

20. Drawer front is added on to the assembled boxed drawer. It is secured with glue and screws.

Finishing

Secure the back panel to the chest with screws only; glue is not required. When all exposed screw holes have been plugged, give the piece a final sanding.

Oak is an open-grained wood and is usually filled with paste wood filler. However, you can dis-

pense with this step in order to retain the natural look of the wood. For a durable, easy-to-apply finish, try clear Deftco Danish Oil Finish or a similar product. This is a blend of tung oil and polyurethane resin that penetrates deeply into the wood. It seals and hardens the fibers and brings out the warmth and natural beauty of the wood. Dust and run problems usually encountered with top-coat finishes are nonexistent with this one. You simply apply a liberal coat and allow it to soak into the wood for 30 minutes. The surplus is then wiped off, and a second coat is applied. This is also followed by wiping after a soaking period.

21. This finish, a penetrating resin, is easy to apply, enhances the color of the wood, and provides tough protection.

Aluminum Pan

The pan for the rain shoes is made by folding up four sides in thin aluminum sheet. Draw the layout for the cutting and folding lines as indicated in the plan. Use a metal straightedge and a sharp utility knife to cut the aluminum to size. Cut out slivers about $1/16$ inch wide at the corners to allow bending clearance, then clamp a strip of wood on the fold line with the area to be folded overhanging the work table. Press against the overhang with a piece of wood to form a right-angle bend. When both ends have been bent clamp a deep guide strip along the side fold line(s) in order to clear the already-bent ends. Make the bends, then fold the end tabs around the corners. Use hot-melt glue to seal the corner joints. Also, lay a bead under the tabs to secure them.

22. Sharp-corner bends are made in the aluminum sheet by pressing a board against the side overhanging the worktable. A back-up board is clamped on the fold line.

23. Hot-melt glue makes a good watertight seal at the corners. It is also used to secure the wraparound end tabs.

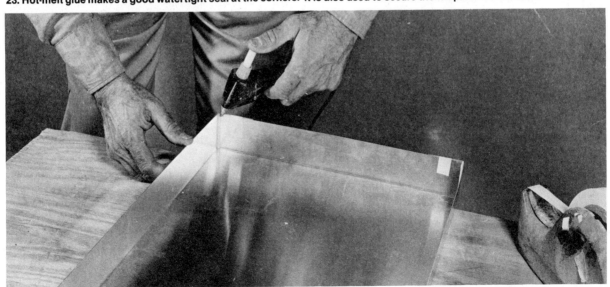

Stained Glass

It should be noted that opalescent rather than transparent stained glass is used for the decorative insert. The reason is that clear stained glass requires rear illumination in order for the colors to be suitably visible, whereas opalescent glass will show a reasonable degree of its colors by light reflected from its front surface. Since the glass panel backs up to a wall in this application there can be no rear illumination.

Although the assembly of a stained-glass panel entails relatively simple soldering of the lead came butt joints, the successful cutting of the glass is not a simple matter for the beginner. This takes much practice, particularly in cutting the curved sections. Since opalescent glass is relatively costly it may prove more economical to have the glass—as well as the leading—cut to size by a professional. Or you may want to go one step further and have the complete panel made to order. You can find stained-glass supply dealers and craftsmen in the Yellow Pages under Glass-Stained & Leaded. Whether you plan to fabricate the panel yourself or to have it partially or completely made up, it will be necessary to draw a full-size pattern of the panel.

The hardware for this project including the porcelain and brass Barber Shop Hooks and Drawer Pulls and a locking Lid Stay may be obtained from Armor Products, Box 290, Deer Park, NY 11729. Write for Hall Butler Hardware Kit price.

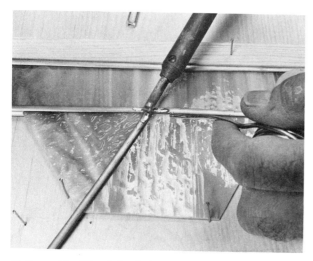

24. Assembly of the stained-glass panel involves soldering the butt joints of the leading strips. Notice how nails are used to hold the glass in place.

25. Glass panel is secured with plastic mirror-retaining clips. Wood cross strips are used to retain the mirror, but the clips could be used here as well.

23. FINGER-JOINT COFFEE MILL

This coffee mill is designed to accommodate a mechanism available from woodworking supply houses (see materials list). The mechanism is easily installed with two screws.

The project will give you the opportunity to learn how to make the finger joint or, as it is sometimes called, the box joint. The joint is used here primarily for visual appeal because the exposed joint makes an effective design element, but the main feature of this joint is strength. The increased glue-contact area provided by the fingers results in strong, lasting assemblies.

Learn the technique of the finger joint by constructing this practical project.

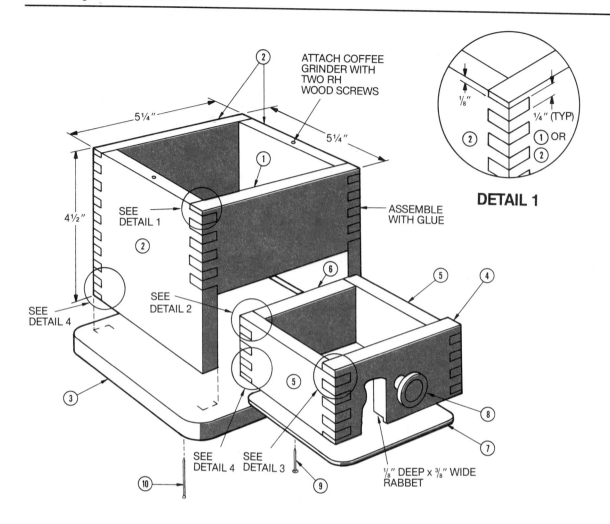

ATTACH COFFEE GRINDER WITH TWO RH WOOD SCREWS

5¼"

5¼"

4½"

ASSEMBLE WITH GLUE

SEE DETAIL 1

SEE DETAIL 2

SEE DETAIL 4

SEE DETAIL 4

SEE DETAIL 3

⅛" DEEP x ⅜" WIDE RABBET

DETAIL 1

⅛"

¼" (TYP)

DETAIL 2

¼"

¼" (TYP)

DETAIL 3

¼"

¼" (TYP)

DETAIL 4
(TYP OF ALL BOTTOM JOINTS)

¼" (TYP)

⅛"

The joint is not difficult to make if you use a simple jig on the table saw. This jig will enable you to produce accurately spaced notches and fingers of uniform width. Once you learn the procedure you can apply it to a great variety of constructions—drawers, chests, boxes, and the like.

To begin, cut all the stock to size. Replace the saw blade with a dado head set up for a $1/4$-inch-wide and $1/2$-inch-deep cut. Note that the depth of the cut is equal to the thickness of the stock. In projects where stock of a different thickness is used the depth of the cut is adjusted accordingly. To make the jig, attach a fresh piece of wood to the miter gauge with screws, then make a pass to cut a notch into it.

Apply a strip of masking tape to the table, positioned so it bridges the notch in the fence. Accurately mark the tape to indicate the cutting path. Detach the fence from the miter gauge, then draw a pencil mark exactly $1/4$ inch away from the edge of the notch toward the end opposite the miter gauge. Position this mark on the tape mark so you can make a second notch cut spaced $1/4$ inch from the first one. Make the cut, then erase the pencil mark.

Turn the fence bottom side up and glue a $1/4$-by-$1/2$-inch block into the second notch. This block should be long enough so it projects a little more than twice the thickness of the workpiece. Use a square to draw two pencil lines accurately: one

FINGER-JOINT COFFEE MILL

MATERIALS LIST

ITEM	QTY	DESCRIPTION
1	1	$1/2'' \times 2^3/8'' \times 5^1/4''$ cherry
2	3	$1/2'' \times 4^1/2'' \times 5^1/4''$ cherry
3	1	$1/2'' \times 6^1/2'' \times 6^1/2''$ cherry
4	1	$1/2'' \times 2^1/8'' \times 4^3/16''$ cherry
5	2	$1/2'' \times 1^7/8'' \times 4^3/4''$ cherry
6	1	$1/2'' \times 1^7/8'' \times 4^3/16''$ cherry
7	1	$1/8'' \times 3^{15}/16'' \times 4^1/2''$ plywood
8	1	drawer knob
9	4	$1/2''$ nail
10	4	$1^1/2''$ finishing nail

ADDITIONAL MATERIALS

Slow-setting hide glue; coffee-mill mechanism; clear finish

Optional: spray enamel; gold paint; paint thinner

Note: The coffee-mill mechanism is available from Constantine's, 2050 Eastchester Road, Bronx, NY 10461, and Armor Products, PO Box 290, Deer Park, NY 11729.

1. Attach a fresh wood fence to the miter gauge, make a pass over the dado head to cut a notch, then mark the notch location on a strip of tape.

2. Detach the fence and shift it $1/4$ inch to the right, using the tape mark for a guide. Use tape to hold the fence to the miter gauge temporarily.

through the center of the first notch and a second line centered between each notch. Attach the fence to the miter gauge so it is in its original position. This completes the setup of the jig.

Make test cuts in two scraps of wood butted face to face. Repeatedly place the previously cut notch over the guide block to make a series of notches. If the test pieces mate properly you can proceed with the work. If not, the guide block is not correctly located and must be readjusted.

3. Cut a second notch in the fence.

4. Glue a guide block into the second notch. It must be sized exactly the same as the notch in both thickness and width.

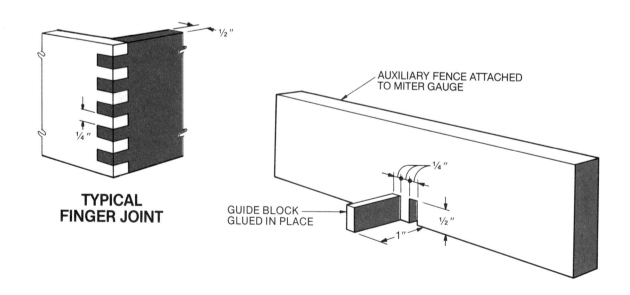

TYPICAL FINGER JOINT

AUXILIARY FENCE ATTACHED TO MITER GAUGE

GUIDE BLOCK GLUED IN PLACE

5. Use a scrap finger from a test piece as a guide to offset a back and side member ¼ inch. Apply tape across the top edges to keep the two pieces in alignment.

6. Align the edge to the centerline nearest the guide block. Make a pass to cut the first notch. A clamp is needed only for the first pass.

7. Place the notch in the work over the guide block to position it for the next cut. A strip of tape in the corner helps to keep the two pieces in close contact.

8. Repeat the step until all the notches have been cut.

Break off a finger from the test piece to use as a spacer. Hold a back and side member together, off-set side to side by the thickness of the spacer. Use a few strips of tape to keep the mated pieces from shifting. Place them against the fence.

Align the edge of the projecting back member with the centerline between the notches in the fence, then clamp the work to the fence. Make a pass to cut a notch in both pieces simultaneously. Remove the clamp; the remainder of the cuts are made without clamping because the guide block will prevent the work from shifting laterally. Note: by starting on the centerline between the notches a half notch (and finger) is formed. This is optional. To start with a full notch, you would simply disregard the line and snug the work up to the guide block to make the first cut.

Place the previously cut notch over the guide block to position the work for the next cut. Continue until all the notches have been cut.

Detach the side member from the back member. Tape the second side member to the back to make the second set of cuts. This time the offset position will be reversed—the member *away* from the fence will extend. Therefore, align the edge of the set-back member with the other centerline, the one directly over the notch in the fence.

Repeat the procedure with the front piece, alternately taped to one side member, then to the other. Here you must be sure to cut notches only to span the width of the front member because the edge below must remain plain.

Since the glue must be applied to so many surfaces, a slow-setting hide glue is recommended for assembling the joints. A quick-setting type such as white glue should be avoided because it may begin to set before you have a chance to complete the application and apply clamp pressure.

The bottom of the drawer is rabbeted 1/8 inch deep by 3/8 inch wide to receive the 1/8-inch-thick plywood panel. Use a router with a 3/8-inch rabbeting bit for this cut. The resultant rounded corners can be squared with a chisel, but this is optional; you can leave the inside corners round and simply round the corners of the panel to match.

Sand the box and base panel before assembly, then apply several coats of clear finish so the finger joints will be clearly visible. Cherry wood is indicated in the plan, but any hardwood species can be substituted.

9. Tape the front piece to a side member to ensure the proper offset alignment. The side member, already cut, goes along for the ride as well.

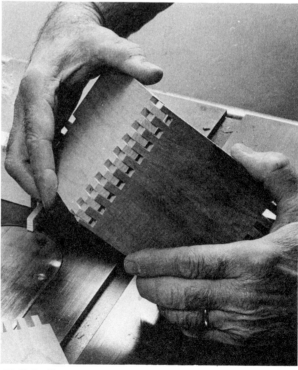

10. If the jig was properly set up, all the parts will join with a slip fit. Forcing should not be necessary.

11. Check all parts before final assembly.

12. Slow-setting hide glue allows ample time for application and clamping.

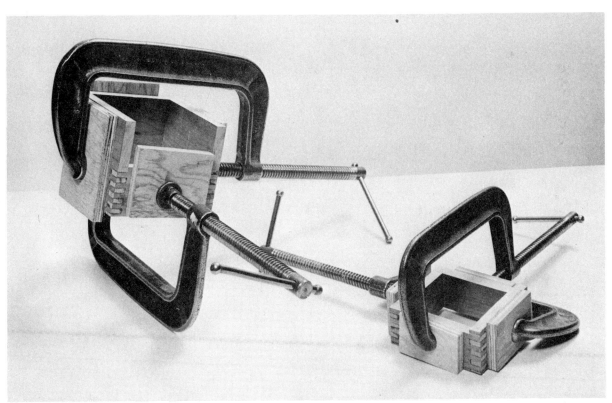

13. Use protective wood pads and two clamps for each assembly. Check for squareness; an off-center clamp axis can twist the assembly.

14. Use a rabbeting bit to cut out the recess for the entire drawer bottom panel.

15. Round the corners of the panel to match the rabbet. Attach with four nails and glue.

16. A clear finish allows the finger joints to show through; staining would subdue the effect.

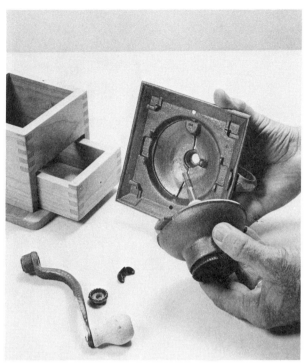

17. Assemble the grinding mechanism with a wing nut and attach it to the box with two screws.

The coffee-grinding mechanism is made of cast iron. The iron has a gray-black color and can be used as is, but for an interesting departure you may opt to give it an antique-white finish as shown.

To obtain this finish you will need a can of Krylon spray enamel Antique White # 1503, a bottle of oil-based gold paint, a small paintbrush, a soft cloth, and some paint thinner.

Apply several light coats of the white enamel to the cast iron and allow each coat to dry thoroughly. Follow with a brushed-on coat of the gold paint. Do *not* allow the gold paint to dry. Soon after it has been applied, lightly dampen the cloth with paint thinner; roll it into a small, tight pad; and wipe off the gold paint from the high spots of the relief design. Treat the wood handle and the drawer knob in the same manner. The result will be a warm white highlighting nicely complemented by gold in the recessed areas.

It should be noted that the lacquer-based spray enamel will remain intact as long as an oil-based paint is applied over it. A lacquer-based gold paint would soften the first coating and the white would rub off during wiping.

18. Apply several light coats of the antique white base color after the bare metal has been prime-coated.

19. Brush gold paint over the entire surface.

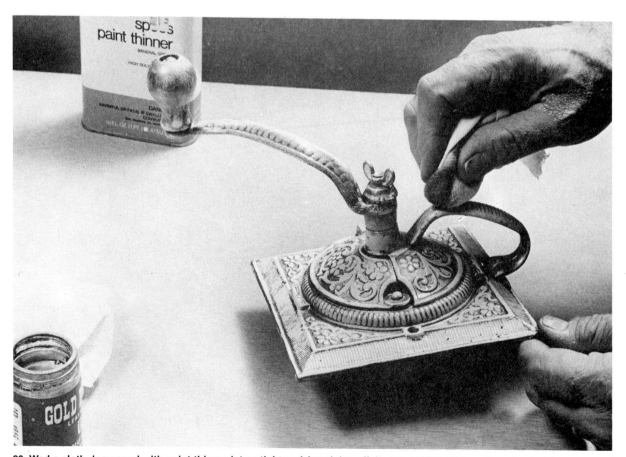

20. Wad a cloth dampened with paint thinner into a tight pad for wiping off the wet gold paint from the raised surfaces.

24. SCULPTED DOLPHIN CLOCK

This graceful leaping dolphin looks like the type of project that might be attempted only by an experienced woodcarver. But this is not so because short-cuts have been purposely devised and incorporated in the design in order to enable the beginner to turn out a fine piece of work on the first attempt.

A novel technique is employed for sculpting the tail and fins; they are partially pre-formed and are added on after the main body has been mostly shaped. This simplifies the task considerably, particularly in view of the fact that the lower fin is angled outward and the tail lies in the horizontal plane. This procedure minimizes labor and lumber cost as well. If the design required the carving from a single block, the job would most likely have to be reserved for the master woodcarver.

This project is not difficult when you tackle the carving in several stages.

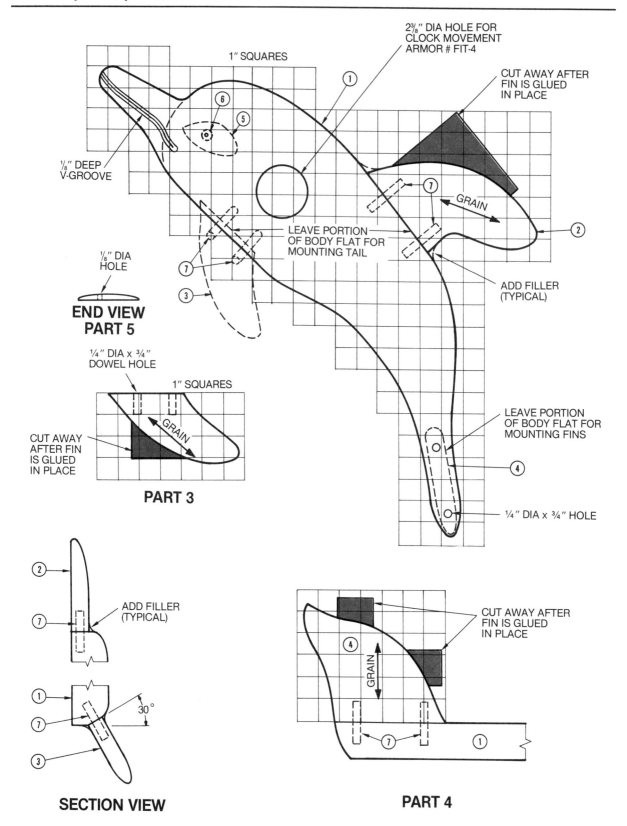

2³⁄₈″ DIA HOLE FOR
CLOCK MOVEMENT
ARMOR # FIT-4

CUT AWAY AFTER
FIN IS GLUED
IN PLACE

1″ SQUARES

⅛″ DEEP
V-GROOVE

⅛″ DIA
HOLE

**END VIEW
PART 5**

¼″ DIA x ¾″
DOWEL HOLE

1″ SQUARES

GRAIN

CUT AWAY
AFTER FIN
IS GLUED
IN PLACE

PART 3

LEAVE PORTION
OF BODY FLAT FOR
MOUNTING TAIL

GRAIN

ADD FILLER
(TYPICAL)

LEAVE PORTION
OF BODY FLAT FOR
MOUNTING FINS

¼″ DIA x ¾″ HOLE

ADD FILLER
(TYPICAL)

30°

SECTION VIEW

CUT AWAY AFTER
FIN IS GLUED
IN PLACE

GRAIN

PART 4

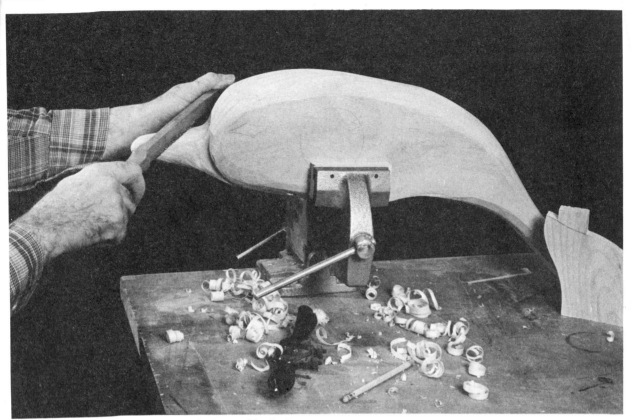

2. Surform is used to shape the nose further and for final contour.

facilitate subsequent gluing of those parts. Outline the flat areas; $3/4$ by $4^1/2$ inches for the fin and $3/4$ by $5^1/2$ inches for the tail. The lower fin also attaches to a flat, but on an angled plane, therefore the outline is not marked until a bevel is cut.

Use a sabre or band saw to cut the outline. Save the cutoff from the lower back to use as a clamp pad when gluing the upper fin. When this has been done, plane a 30-degree bevel at the location of the lower fin. Use a T bevel to check the angle of the bevel. Mark the outline for this fin $3/4$ by $3^1/2$ inches. Also mark an outline $1^1/2$ by 3 inches at the eye-pad location; this too must remain flat.

Work with the spokeshave to contour the front edges of the body and use both the spokeshave and the Surforms to shape the nose. The narrow, round Surform is useful for forming the slight concave in the area near the rear of the mouth. When this has been done draw the outlines for the fins and tail on the $4/4$ stock; include the clamping tabs as they are shown in the plan. They provide a firm grip for the clamp jaws during the gluing operation and are cut off afterward. Make the cuts.

Bore two $1/4$-inch-diameter holes for dowels in

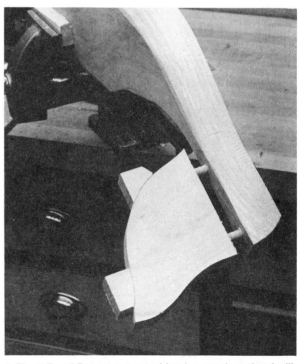

3. The tail and fins are cut out with tab extensions that facilitate clamping. Two dowels are used to strengthen the joints.

The bulk of the sculpting is done with hand tools, including the Surform in several shapes, and the spokeshave. A piece of $^8/_4$-sugar pine is used for the body; $^4/_4$ stock is used for the tail and fins. Sugar pine excels for the purpose because it tools, sands, and takes finish very well.

To begin, draw the outline of the body on the board and indicate clearly the location of the upper fin and tail. These areas must be left flat in order to

SCULPTED DOLPHIN CLOCK

MATERIALS LIST

ITEM	QTY	DESCRIPTION
1	1	$1^3/_4'' \times 7^1/_2'' \times 27^{15}/_{16}''$ pine ($^8/_4$ stock)
2	1	$^3/_4'' \times 5'' \times 8''$ pine
3	1	$^3/_4'' \times 3'' \times 6^1/_2''$ pine
4	1	$^3/_4'' \times 5^3/_4'' \times 6^5/_8''$ pine
5	1	$^1/_4'' \times 1^3/_8'' \times 2^3/_4''$ pine
6	1	plastic eye # KA
7	6	$^1/_4'' \times 1^1/_2''$ dowels

ADDITIONAL MATERIALS

Clock movement # FIT-4; wood-filling compound; stain; finish

Note: The clock movement and plastic eye are available from Armor Products, PO Box 290, Deer Park, NY 11729.

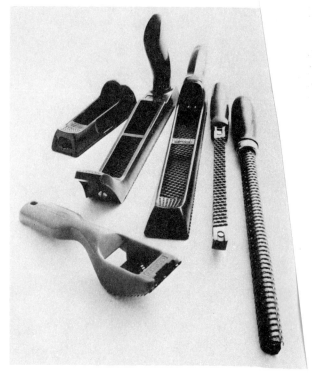

Surform tools are available in a variety of sizes and shapes.

1. Deep shavings are made with the spokeshave to remove much of the bulk. Surform is used to form the cheek.

the bottom edges of the fins and tail. Use dowel centers to transfer the center marks for matching holes in the body accurately. Insert a center in each hole, then align the part over the mating surface and press together to form indents. Bore the holes for the dowels in the body.

The upper fin is clamped by using the scrap from the lower body as a clamp jaw pad. To facilitate clamping the angled lower fin, temporarily nail a small block of wood with a beveled surface to the back of the body, positioned to receive the clamp jaw. When the glue has set, remove the clamp tabs with a sabre saw. Shape the fins and tail. The small one-hand Surform tool with the curved blade is well suited for this purpose.

The sharp corners where the fins and tail join the body are packed with wood-filling compound to produce blending fillets. Apply a sufficient quantity of filler and shape it with the index finger or thumb. Sand thoroughly to smooth out all tool marks.

The line for the mouth is made with a V-shaped carving chisel or with a utility knife. It should be cut to a depth of about $1/8$ inch. The eye pad is shaped from a piece of $1/4$-inch stock. Bore a $5/32$-inch hole $1/2$ inch deep after the eye pad has been glued in place. This hole will accept a plastic eye. The eye and the clock movement are available by mail-order (see the Materials List).

Bore a $3/8$-inch hole 1 inch deep in the rear to permit hanging. Bore a $2 3/8$-inch through hole for the clock movement. Apply stain and clear finish or paint as desired *before* installing the eye and clock.

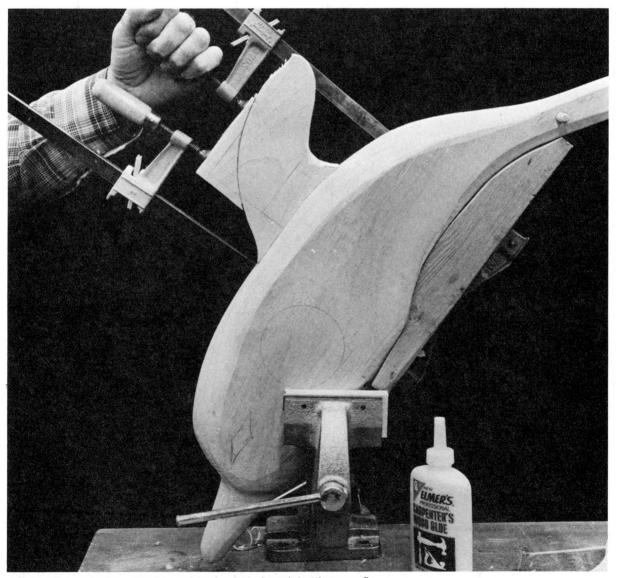

4. Curved lower cutoff is utilized as a clamping pad when gluing the upper fin.

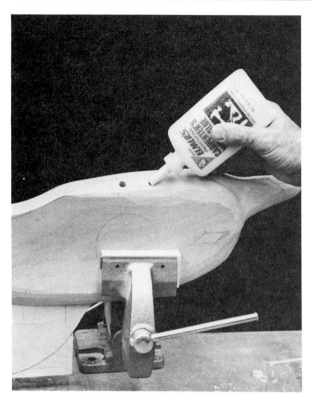

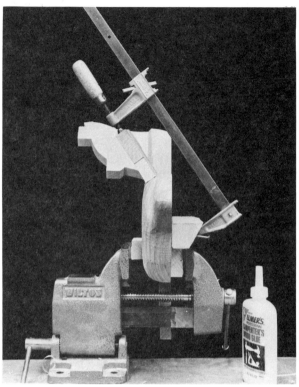

5. It is important to leave a flat area at the locations where the extremities are to be glued. This flat for the lower fin is angled.

6. A block is temporarily nailed to the back of the figure to provide a clamping surface opposite the angled fin.

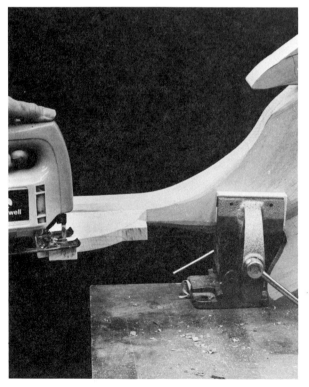

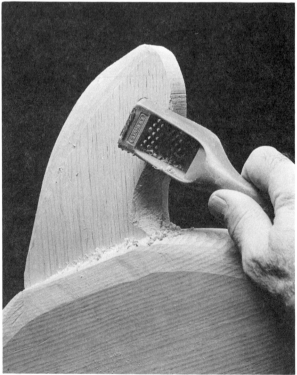

7. Clamping pads are cut off after the glue has set.

8. Curved Surform is used in tight quarters where the spokeshave cannot be manipulated.

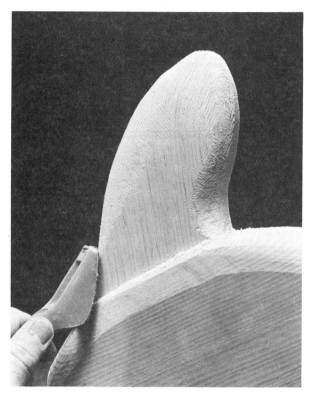

9. Almost completed fin. Thorough sanding follows to remove all tool marks.

10. Curved-back spokeshave is especially useful for working small inside curves.

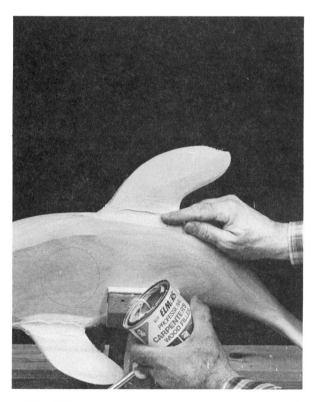

11. Wood-filling compound is used to form fillets to blend inside corners to the surrounding planes.

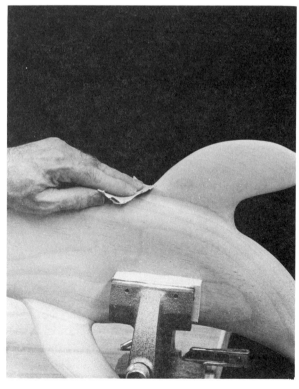

12. Allow the filler to dry thoroughly before sanding. The hole for the clock movement is bored next.

25. PICKET FENCE HOW-TO

Ordinary, run-of-the-mill fences can be purchased at prices that are relatively within reason; but the costs for custom, made-to-order fences can run prohibitively high. If you want an affordable fence of distinctive design you might consider building it yourself.

Shown here are some interesting, out-of-the-ordinary fences that you can build. Offhand, they may seem to be too involved and difficult to duplicate, but this is not so. If you follow the construction shortcuts described you will discover that any of these fences can be made with relative ease and a minimum of tedious labor. You can of course modify the designs to suit your own taste and simply apply the basics.

One of the most important considerations in the construction of a fence is the choice of wood. Unless the wood you use for soil-contact is resistant to decay and insect attack, all your efforts and expenditures may be in vain. The heartwood of redwood, cypress, and cedar are especially suitable for soil-contact fence members. However, with the exception of cedar, these species are very costly and possibly not readily available in the required dimensions. Therefore your best approach would be to use pressure-preservative-treated lumber, especially for the in-ground and soil-contact members. Since the cost differential between regular and treated lumber is not too great, it would be wise to use treated stock for as much of the fence construc-

tion as possible. If you opt to use regular lumber for the above-ground members, a thorough coating of brush-applied wood preservative used according to the manufacturer's instructions will prove well worth the effort. For best results the preservative should be applied after the parts have been cut to size but before assembly.

In the designs shown, with the exception of the square picket fence, the corner posts are made of 6×6 lumber faced with $3/4$-inch stock. The rails are let into through mortises or notches cut in the inside face boards. This is an important design feature because it permits shop construction of all the joint components up to the stage of sectional assembly. The separate post installation procedure makes the job easier and results in foolproof final assembly.

The round picket fence is made up of 8-foot sections between posts while the square picket fence is made up of 4-foot sections. The other two fences can run continuously between corner posts but utilize 4-by-4-inch intermediate rail support posts on 4-foot centers throughout the run. The basic setting procedure is as follows: dig the holes for the corner and intermediate posts and drop the posts loosely in place. For the round and square picket fences pre-assemble each picket section. Attach the inside face boards to each section of the round picket fence. For the other two fences make up the rail sections and attach the inside face boards. Be certain to insert all the rails so the ends are flush with the

282

Custom fences are not difficult to make when shortcut methods are used.

backs of the face boards. Drive nails into each edge to keep the parts in place.

Position and prop the assembled fence sections or the bare rail sections in place, level, and plumb, checking with a carpenter's level. Attach the inside face boards to the corner posts and nail through the upper rails into the tops of all intermediate posts (where applicable). Nail temporary diagonal braces to keep the parts in place, then back-fill the post holes with soil; for a better installation use cement.

Follow the final assembly methods described below for each particular fence. Notice that the photographs show short lengths of posts. This was

done for convenience in photography. In an actual installation the posts must be long enough to be set to a depth about a foot below the frost line. This will vary depending upon your geographical location. You can obtain this data from your local building department. Also check with them concerning any set-back restrictions for fences along a property line.

Several pointers: to ensure good drainage, bottom each post hole with about 6 inches of gravel before setting the post. Also pour in a few inches of gravel before filling the hole with cement. Use only hot-dipped galvanized or aluminum nails for all assembly.

Arch Spaced Picket

This fence is supported at 4-foot intervals with 4×4 intermediate posts. The lower rail is attached to the post by means of a half-lap joint. This is simply a notch cut half way through each member so they will nest for a flush fit. The upper rail rests directly on the top of the intermediate posts. A skirt, which appears to form the base for the fence, is actually suspended 1 inch above the ground. It is nailed to a short length of 4×4 at the base of the corner post and to the lower face of each intermediate post.

Use the radial-arm saw and make repeated kerf cuts to form the notches in the rails and posts. Note: for this fence and the others that require notching, the job can be done with a portable saw if a radial-arm saw is not available.

Cut two ³/₄-inch boards to size to make the inside and outside end post facing members. These should be about ¹/₈ inch wider than the thickness of the post. Cut the front and back facing boards 1¹/₂ inches wider overall. The rails are made of 2×4 stock. Lay out the two through mortises on the inside face board, then bore two 1¹/₂-inch-diameter

MATERIALS LIST		
CORNER POST		
ITEM	QTY	DESCRIPTION
1	1	$6 \times 6 \times$ length (as required)
2	2	³/₄″ × 7¹/₈″ × 42″ face board
3	1	³/₄″ × 5⁵/₈″ × 42″ face board (inside)
4	1	³/₄″ × 5⁵/₈″ × 42″ face board (outside)
5	1	³/₄″ × 6″ × 6″ post cap
6	1	³/₄″ × 11″ × 11″ post cap
7	4	³/₄″ × 2¹/₂″ × 8⁵/₈″ trim
8	4	³/₄″ × 10¹/₈″ cove molding
9	4	³/₄″ × 8⁵/₈″ cove molding
10	1	³/₄″ × 5″ × 28″ trim
FENCE		
11		2×4 upper rail
12		2×4 lower rail
13		4×4 intermediate post
14		³/₄″ × 2¹/₂″ × 25¹/₄″ pickets
15		³/₄″ × 2¹/₂″ × 3¹/₂″ upper arch pickets
16		³/₄″ × 2¹/₂″ × 5″ lower arch pickets
17		³/₄″ × 1¹/₂″ skirt cap
18		³/₄″ × 5″ skirt
19		³/₄″ half round molding
20		³/₄″ cove molding
21		³/₄″ × 2″ trim
22		³/₄″ × 6³/₄″ fence top
23		$4 \times 4 \times 8³/₄″$ post
24		³/₄″ × 1¹/₂″ trim

Note: the actual dimensions of 6×6 stock are 5¹/₂ × 5¹/₂ inches.

1. A hole saw and table saw team up to mass-produce the arched members of this fence.

holes tangent to the ends of the rectangles and use a sabre saw to make the cutouts to receive the ends of the rails. Next, cut a length of 4×4 equal to the dimension from the bottom of the lower opening to the bottom edge of the face board. Nail this block in place; it serves as a nailing surface for the skirt. Insert the rails into the end face boards for the number of end posts that you have. Drive a nail through the edges and into the rails.

Take the rail sections to the site and attach them to the posts, then set the posts permanently. Note: for this fence and the Gingerbread Picket (page 293), where intermediate rail support posts are required at 4-foot intervals, each alternate post should be located on 8-foot centers to provide nailing surfaces for the butting rail ends. This assumes the use of 8-foot-long rails. You can use any size rail (12, 16, or 20 feet) provided the rails are straight and warp-free. Whatever size rail you use, simply

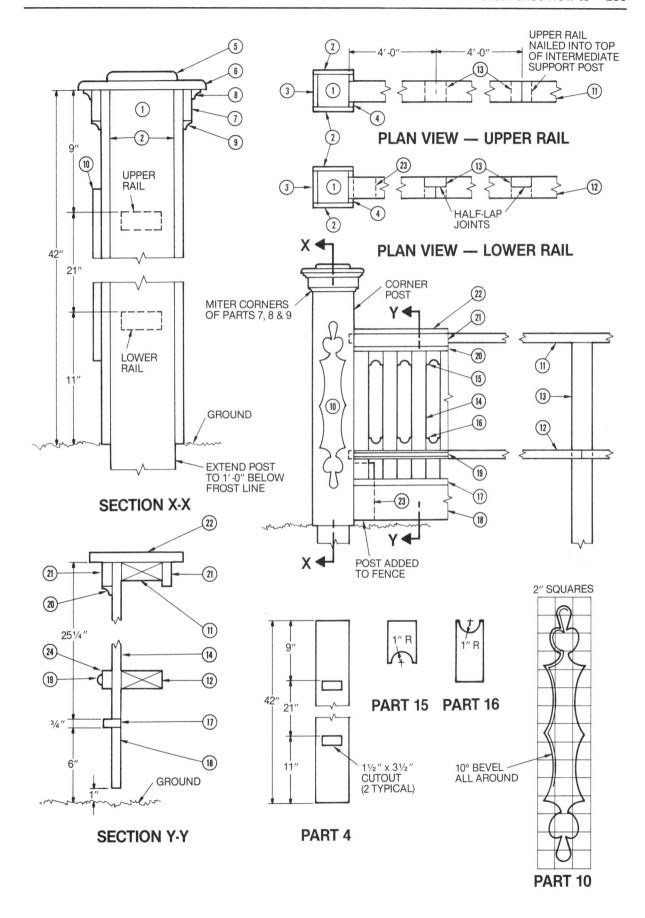

SECTION X-X

PLAN VIEW — UPPER RAIL

PLAN VIEW — LOWER RAIL

HALF-LAP JOINTS

UPPER RAIL NAILED INTO TOP OF INTERMEDIATE SUPPORT POST

9″

42″

21″

11″

UPPER RAIL

LOWER RAIL

GROUND

EXTEND POST TO 1′-0″ BELOW FROST LINE

4′-0″   4′-0″

MITER CORNERS OF PARTS 7, 8 & 9

CORNER POST

POST ADDED TO FENCE

SECTION Y-Y

25¼″

¾″

6″

1″

GROUND

PART 4

9″

42″   21″

11″

1½″ x 3½″ CUTOUT (2 TYPICAL)

PART 15   PART 16

1″ R

1″ R

2″ SQUARES

10° BEVEL ALL AROUND

PART 10

2. This is how the basic frame will appear when it is set up at the site. The facing boards wrap around the 6 × 6 corner post to conceal it.

3. The skirt is propped up so it will be nailed 1 inch above ground level.

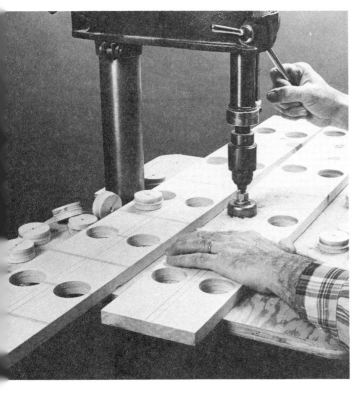

4. Lay out the centers and cutting lines, then use a hole saw to cut the circular openings. Cut from both sides.

5. Rip through the center of each board to produce strips 2½ inches wide. Notice that double lines are used in the layout to compensate for saw kerf waste.

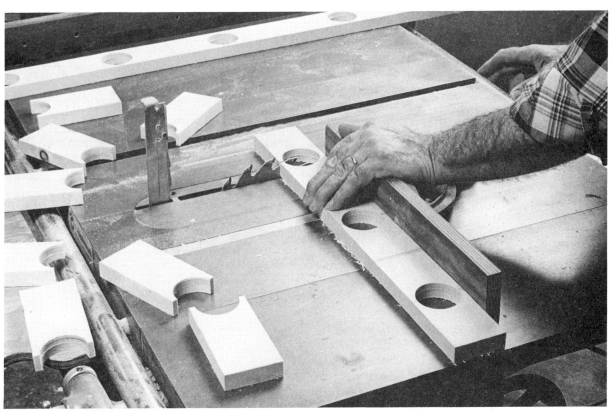

6. Crosscuts through the centers of the holes produce ready-to-assemble picket dividers.

7. Use one nail at each rail to secure the pickets and dividers. This phase is done at the site after the posts and rails have been set up.

make sure that a support post will center directly under the butting ends.

Cut the stock for the skirt to size, nail on the cap strip, then nail the skirt in place. The pickets and trim are then added.

The arched picket spacers are easily mass-produced by cutting a series of appropriately spaced holes in ³/₄-inch stock and then cutting them apart by ripping and crosscutting. Use a 2-inch hole saw and cut about two thirds through the face of the stock. This will allow the pilot drill bit to penetrate the back surface. Flip the workpiece over and continue cutting to complete the holes. Working from both sides ensures clean-cut edges and prevents the waste from jamming in the saw.

The fancy appliqué is made by setting the band or scroll saw for a 10-degree bevel cut. Due to the configuration it will be necessary to tilt the saw table alternately left and right. You will notice that after the outline cuts are completed the waste around the small curves at the ends will not be severed completely. Use a chisel here.

Nail the front and back face boards in place to box in the post, attach the appliqué, then add the cap and trim moldings. The cap trim pieces are assembled with 45-degree miters.

Square Picket

The pickets for this fence are made by ripping square strips out of $^6/_4$ stock. This lumber is usually dressed down to a thickness of $1^3/_8$ inch, plus or minus a little. Check this out and set your saw fence accordingly.

The pickets are set flush into through notches cut into the backs of the upper and center 2×3 rails. The notches in the lower rail are only $^3/_4$ inch deep. To cut numerous partial notches such as these with a chisel would be a monumental task, but there is a shortcut; two pieces of $^3/_4$-inch-thick stock are used to make up the bottom rail. Through notches are cut into one piece only. When the second piece is added to the bottom the partial notches will be automatically formed.

Accurate vertical alignment of the notches is essential for trouble-free installation of the pickets. This is easily accomplished by ganging the three rail members and cutting the notches in each piece simultaneously. Drive a temporary screw into each end to keep the pieces from shifting. Mark the pack for the notch locations, then use the radial-arm saw to make repeated kerf cuts to form the notches to the depth and width of the pickets.

A simple procedure—which utilizes a new fence section—will enable you to align the work for the

MATERIALS LIST		
CORNER POST		
ITEM	QTY	DESCRIPTION
1		$4 \times 4 \times$ length (as required) corner post
2		$4 \times 4 \times$ length (as required) intermediate post
3		4″ finial
FENCE		
4		$2 \times 4 \times 48^1/_2$″ upper rail
5		$2 \times 4 \times 48^1/_2$″ middle rail
6		$^3/_4$″ $\times$ $3^1/_2$″ $\times$ $48^1/_2$″ (2 required)
7		$1^3/_8$″ $\times$ $1^3/_8$″ $\times$ 21″ pickets
8		$1^3/_8$″ $\times$ $1^3/_8$″ $\times$ 36″ pickets
9		$2^1/_2$″ chair rail molding (cut to $2^1/_4$″)
10		$2^1/_2$″ chair rail molding (cut to $1^1/_2$″)
11		$^3/_4$″ $\times$ $1^1/_2$″ rear rail trim

Note: the actual dimensions of 4×4 stock are $3^1/_2 \times 3^1/_2$ inches.

numerous notch (dado) cuts on the radial-arm saw with ease and speed. See Sizing Notch Cuts (page 140) for details on how to set it up. Use a dado head or a regular saw blade to cut the notches. When all the picket notches have been cut, turn the pack up side down and cut the notches at the ends for the post full-lap joints. Remove the screws, then join the $^3/_4$-inch solid bottom rail to the $^3/_4$-inch notched rail member using $1^1/_2$-inch nails and waterproof glue.

The picket points are also cut in gangs to save time. Lay seven or eight picket pieces side by side, each one offset at the ends in steps $1^3/_8$ inches apart. Hold them together with strips of masking or duct tape. Set the saw for a 45-degree miter cut. Make a pass to cut a bevel that will intersect the center end of each picket. Remove the tape, turn each piece to a new side, retape, and make another cut. Repeat this step until the four cuts are made to form the pyramid points.

You will notice that in cutting the initial notches in the ganged rails, the upper rail will have too many notches; it requires seven but it will have fifteen. This could be avoided by initially cutting the notches in this piece separately; but in the interest of precise alignment, it is better to cut the extra notches and simply plug them with glued-in blocks cut from scrap picket stock.

The notches in the posts that mate with the rails are cut by hand with a chisel. You can save much time and make this task easier if you first bore $^1/_8$-inch-diameter holes into each corner of the cutouts so they intersect within the post. This will simplify trimming the inside corners clean. Also, make two

A number of special tricks are used to speed the construction of this fence.

SQUARE PICKET

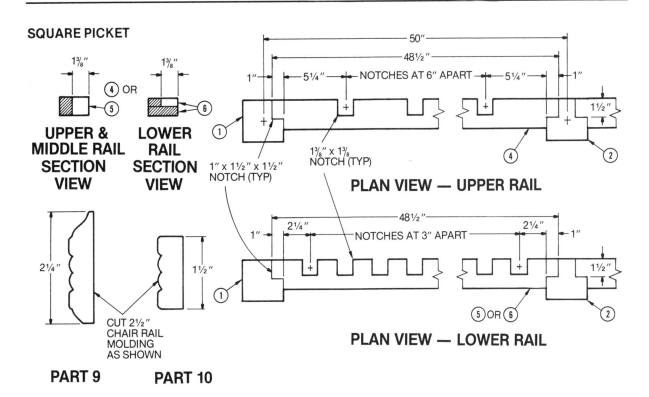

$1\frac{3}{8}''$

(4) OR
(5)

UPPER & MIDDLE RAIL SECTION VIEW

$1\frac{3}{8}''$

(6)

LOWER RAIL SECTION VIEW

$1'' \times 1\frac{1}{2}'' \times 1\frac{1}{2}''$ NOTCH (TYP)

$1\frac{3}{8}'' \times 1\frac{3}{8}''$ NOTCH (TYP)

50''

$48\frac{1}{2}''$

1'' $5\frac{1}{4}''$ NOTCHES AT 6'' APART $5\frac{1}{4}''$ 1''

$1\frac{1}{2}''$

PLAN VIEW — UPPER RAIL

$2\frac{1}{4}''$ $1\frac{1}{2}''$

CUT $2\frac{1}{2}''$ CHAIR RAIL MOLDING AS SHOWN

PART 9   PART 10

$48\frac{1}{2}''$

1'' $2\frac{1}{4}''$ NOTCHES AT 3'' APART $2\frac{1}{4}''$ 1''

$1\frac{1}{2}''$

(5) OR (6)

PLAN VIEW — LOWER RAIL

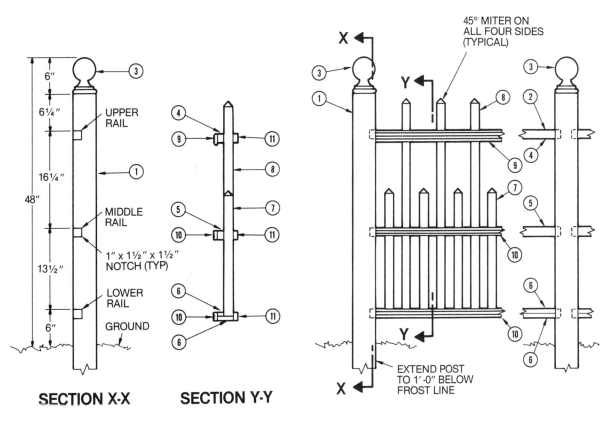

SECTION X-X

6''
UPPER RAIL
$6\frac{1}{4}''$
$16\frac{1}{4}''$
48''
MIDDLE RAIL
$1'' \times 1\frac{1}{2}'' \times 1\frac{1}{2}''$ NOTCH (TYP)
$13\frac{1}{2}''$
LOWER RAIL
GROUND
6''

SECTION Y-Y

45° MITER ON ALL FOUR SIDES (TYPICAL)

EXTEND POST TO 1'-0'' BELOW FROST LINE

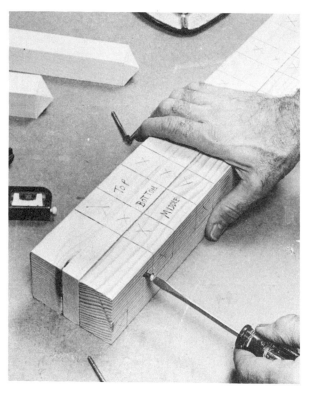

1. Gang three rail members and secure them with a screw at each end for cutting.

2. After one kerf cut is made, slide the pack to align that kerf with the second one in the fence to establish the width of the notch. This trick saves considerable time in alignment.

3. Make repeated cuts between the outer kerfs to remove the waste. Shave off the thin remaining slices with a chisel. A dado head would make even quicker work of this operation.

4. Glue and nail the 3/4-inch notched rail member to a solid 3/4-inch base board: a quick way to mass-produce mortises.

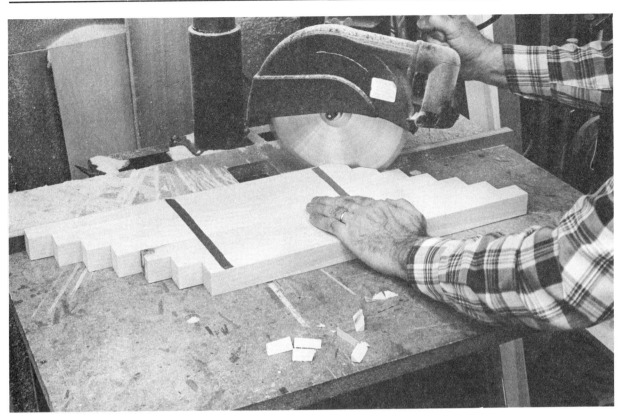

5. Tape the picket squares together in offset steps to speed up the pointing of the tops. Rotate and retape the pickets until the four sides have been mitered.

6. Four mitered saw cuts result in clean-cut points. Saw is set at 45 degrees for this operation.

7. Plug the alternate extra notches in the upper rail with small blocks of picket stock; then glue and nail.

8. Lay the rails in place to mark the mortise locations on the fence posts.

9. Corner holes and partial saw cuts set the limits for carving out the mortises.

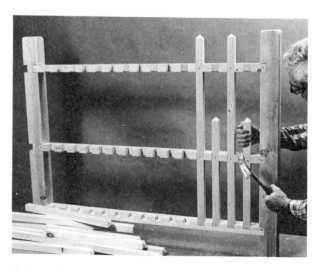

10. Use the chisel to cut the notches in the posts. The predrilled holes speed up the corner squaring.

partial saw cuts into the outside corners of the posts. This will reduce the amount of end-grain chiseling required. The corner posts are notched on one corner only; intermediate posts are notched on two corners.

To assemble this fence, insert the rails into the posts to ensure alignment; then proceed to nail in the pickets. You can make up alternate picket sections, without posts, but do not attach them until the units are brought to the site. Set the completed sections into the holes, prop them in place, and attach the alternate picket sections. Fill the holes when all the sections are combined.

The tops of the posts can be slightly rounded by power sanding. The post finials, available at lumberyards, come with screw dowels. Bore pilot holes in the post tops and screw the finials into place.

The trim used on the rail fronts is $2^1/2$-inch chair rail molding. This is trimmed to a width of $2^1/4$ inches for the upper rail and $1^1/2$ inches for the center and lower rails. Rip from both edges to preserve the molded area.

11. Detail shows how the rails tie in to the posts.

12. One nail at each junction holds the pickets in place.

Gingerbread Picket

This fence utilizes a corner post, intermediate 4 × 4 posts, an upper rail, and a skirt arrangement somewhat similar to the Arched Picket fence. The lower rail is essentially the same as the lower rail of the Square Picket fence; it is partially notched to receive the bases of the pickets.

Begin construction by drawing a full-size pattern of the picket on white paper (use the enlarging-by-squares method). Cut out the pattern and use rubber cement to mount it to a piece of 1/8-inch-thick

Scrolled Victorian styling sets this fence in a class of its own.

hardboard. Carefully saw around the outlines to make a tracing template. The white paper provides good visibility in following the lines on the dark board. It can be removed after cutting and sanding. Trace the picket pattern onto the workpieces.

A series of appropriately sized holes bored into each picket blank will save much time and effort in sawing; the pre-formed curves will eliminate the need for painstaking sawing at these points. Use a circle template guide to locate the center marks for drilling. Use 1/2-inch and 3/4-inch bits for the smaller holes and a 2 1/2-inch hole saw for the large hole.

Make the inside cutouts with a sabre saw; work one picket at a time. When the inside cuts have been made, gang three or four pickets and cut the outside contours simultaneously on the band saw. To ensure positive alignment in ganging the boards, simply insert two 1/2-inch drill bits (or dowels) into the 1/2-inch-diameter holes that were originally drilled in the center of the picket. When the several boards are in alignment partly drive in four finishing nails to hold them together. For best results work with a 3/8-inch-wide band saw blade.

The wide notches for the upper portion of the lower rail can be cut with a dado head on the radial-arm saw. For safety and efficiency, install a temporary high fence to support the work piece on edge.

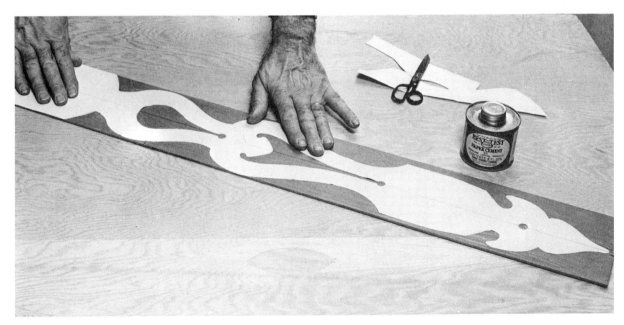

1. Mount the paper picket pattern on hardboard with rubber cement for an easy-to-follow outline in cutting the template.

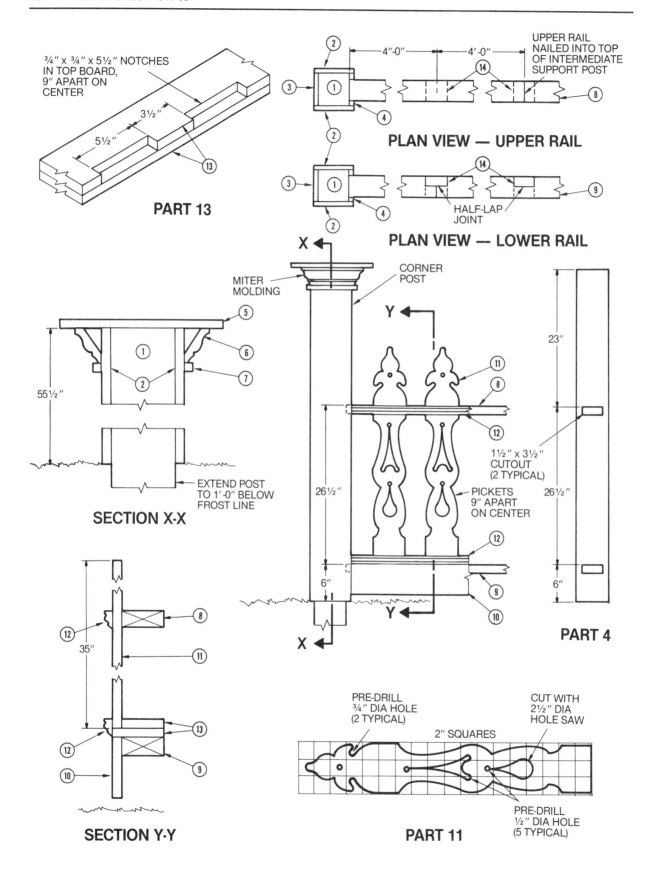

¾" x ¾" x 5½" NOTCHES
IN TOP BOARD,
9" APART ON
CENTER

3½"

5½"

⑬

PART 13

②

4'-0"   4'-0"

UPPER RAIL
NAILED INTO TOP
OF INTERMEDIATE
SUPPORT POST

③   ①

⑭

④

②

⑧

PLAN VIEW — UPPER RAIL

②

③   ①

⑭

④

⑨

②

HALF-LAP
JOINT

PLAN VIEW — LOWER RAIL

X

MITER
MOLDING

CORNER
POST

Y

23"

⑤

①

⑥

②

⑦

⑪

⑧

⑫

55½"

1½" x 3½"
CUTOUT
(2 TYPICAL)

EXTEND POST
TO 1'-0" BELOW
FROST LINE

26½"

PICKETS
9" APART
ON CENTER

26½"

SECTION X-X

6"

⑫

⑨

⑩

Y

X

PART 4

⑫   ⑧

35"

⑪

⑬

⑫

⑨

⑩

SECTION Y-Y

PRE-DRILL
¾" DIA HOLE
(2 TYPICAL)

CUT WITH
2½" DIA
HOLE SAW

2" SQUARES

PRE-DRILL
½" DIA HOLE
(5 TYPICAL)

PART 11

GINGERBREAD PICKET

MATERIALS LIST

CORNER POST

ITEM	QTY	DESCRIPTION
1	1	6 × 6 × length (as required)
2	2	3/4″ × 71/8″ × 551/2″ face board
3	1	3/4″ × 55/8″ × 551/2″ face board (inside)
4	1	3/4″ × 55/8″ × 551/2″ face board (outside)
5	1	3/4″ × 14″ × 14″ post cap
6	4	35/8″ × 111/8″ crown molding
7	4	3/4″ × 3/4″ × 85/8″ trim
FENCE		
8		2 × 4 upper rail
9		2 × 4 lower rail
10		3/4″ × 5″ skirt
11		3/4″ × 51/2″ × 35″ scrolled picket
12		11/2″ base molding

Note: the actual dimensions of 6 × 6 stock are 51/2 × 51/2 inches.

Assemble the lower rail in this sequence: nail the lower unnotched board to a 2 × 4 rail. Note: this piece overhangs the end and front of the 2 × 4 by 3/4 inch. Next, add the notched member. Finally, add the skirt board, which is nailed directly to the edge of the 2 × 4 rail. Turn the piece over and cut the necessary half-lap notches to allow the rail to set into the intermediate support posts.

The inside face board is mortised through to receive the rail ends in the same manner as was described for the Arched Picket fence. Secure the rails to the posts at the site and set the posts permanently before attaching the pickets. O G floor molding is used to trim the pickets. It is mounted in reverse position opposite the upper and lower rails.

The cap of the corner post is set off with 35/8-inch crown molding with mitered corners. True-fitting miters for this type molding need to be cut with a jig. Refer to the sections on Sawing Jigs if you need a refresher on the procedure for making the compound miter cut.

2. The rigid template makes easy work of tracing the cutting lines on the picket blanks.

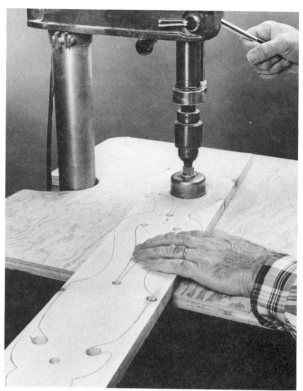

3. Use a circle guide to spot the centers for the holes.

4. Pre-drilled holes save considerable time in sawing the outlines. This also results in perfectly formed inside curves.

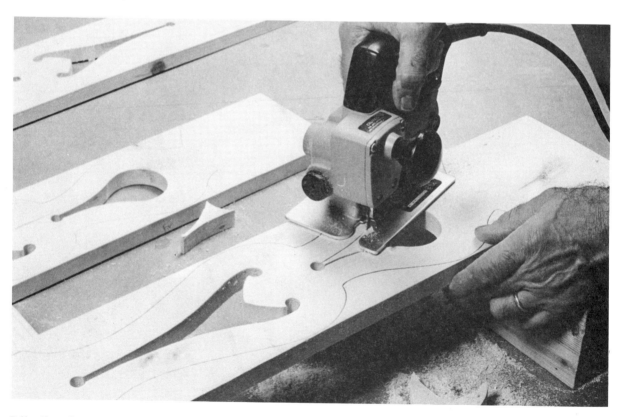

5. Use the sabre saw to make the inside cuts.

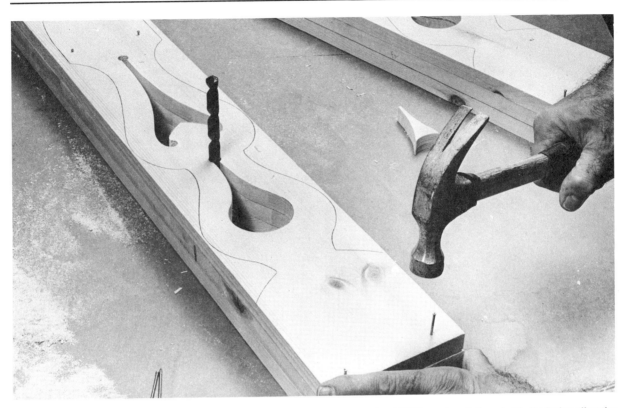

6. After the inside cuts have been made, gang several picket boards and cut them simultaneously on the band saw. Matched-size drill bits or dowels inserted in the holes align the boards for tack nailing.

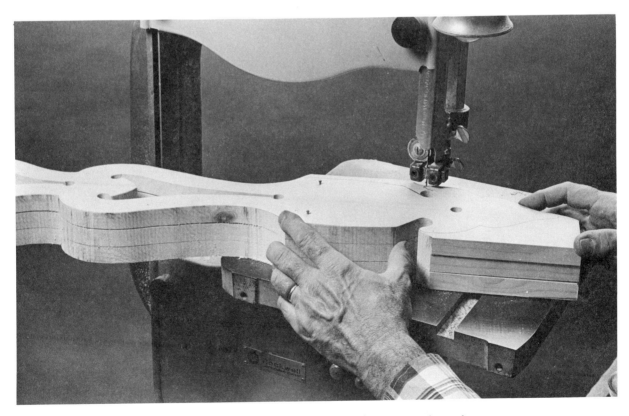

7. When sawing up to the holes take care to keep on the line so you meet the curve on a tangent.

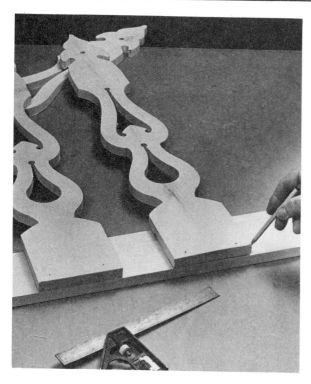

8. Space the pickets and mark the lower rail for the notch cuts.

9. A high fence is a necessity for this operation. A dado head set for a wide cut makes fast work in cutting the notches.

10. Assembly of the lower rail begins by nailing a ¾-inch board to a length of 2 × 4.

11. Notched member follows to produce pre-formed mortises.

12. View of the post as it would be boxed in. This and the following steps would be done at the site after the post is set in the ground and the rails secured.

13. Two hot-dipped galvanized nails secure the pickets. These nails will not rust, and they grip exceptionally well due to their rough surface.

14. Attach trim molding with finishing nails. The nails can be set and the indents filled for a neater job.

15. This is how the crown molding is set in place.

Round Picket

This fence is made with three 2 × 4 rails and lengths of 1⅛-inch-diameter closet pole (round stock over 1 inch in diameter is not called dowel).

As is true with the Square Picket fence, vertical alignment of the pickets is essential if a good job is to result. The poles insert through holes in the center rail and in blind holes in the upper and lower rails. Each 8-foot picket section is supported by a 6 × 6 boxed-in post. The rail-supporting inside face boards are notched from the rear edge, as opposed to the inside cutouts required on the other fences. This is done for appearance. The arrangement places the rails closer to the back of the posts.

The best way to get all the holes in perfect vertical alignment is by stacking and clamping the rails one atop the other and drilling long pilot holes

MATERIALS LIST		
CORNER POST		
ITEM	QTY	DESCRIPTION
1	1	6 × 6 × length (as required)
2	2	¾″ × 7⅛″ × 42″ face board
3	1	¾″ × 5⅝″ × 42″ face board (inside)
4	1	¾″ × 5⅝″ × 42″ face board (outside)
5	1	1½″ × 6″ × 6″ post cap
6	1	1½″ × 11″ × 11″ post cap
7	4	¾″ × 8⅝″ cove molding
FENCE		
8		2 × 4 upper rail
9		2 × 4 middle rail
10		2 × 4 lower rail
11		1⅛″ × 29½″ closet pole pickets

Note: the actual dimensions of 6 × 6 lumber are 5½ × 5½ inches.

The round picket fence is an easy one to build, but precise alignment is essential.

through the bottom rail and center rail and partly into the top rail. Notice that the rails are stacked top, middle, and bottom. The stack is then turned up side down for drilling. Identify each rail to avoid mix ups. Use a drill guide or a drill press to obtain perfectly perpendicular holes. Allow the drill to penetrate only slightly into the bottom face of the top rail (which is at the bottom during drilling).

Although the poles are nominally 1⅛ inches in diameter, they usually measure about 1 1/16 inches. Nevertheless, 1⅛-inch holes should be bored. This allows some leeway in sliding the pickets into place —a necessity for those poles that are not perfectly round on occasion. Bore the blind holes in the bottom of the upper rail and in the top of the lower rail ½ inch deep. Bore through holes in the middle rail. The previously drilled pilot holes will ensure perfectly aligned holes.

Cut the notches in the inside post face boards and use them to space the upper and middle rails while inserting the poles. Remove the face boards temporarily to allow placement of the bottom rail. Check with a large square to see that the assembly is true, then proceed to drive a nail through the backs of the rails and into the sides of the poles. Take the units to the site.

Set the posts at the corners and between each picket section. When the posts have been secured in the ground, complete boxing them in with the facing boards and add the post caps.

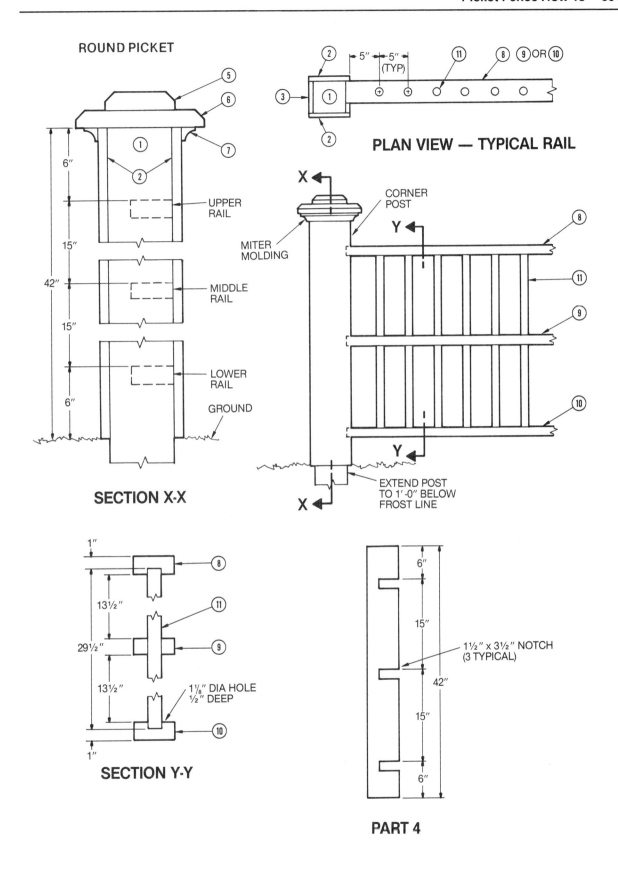

ROUND PICKET

SECTION X-X

UPPER RAIL

MIDDLE RAIL

LOWER RAIL

GROUND

6"
15"
42"
15"
6"

PLAN VIEW — TYPICAL RAIL

5"   5"
(TYP)

CORNER POST

MITER MOLDING

EXTEND POST TO 1'-0" BELOW FROST LINE

SECTION Y-Y

1"
13½"
29½"
13½"
1"

1⅛" DIA HOLE ½" DEEP

PART 4

6"
15"
42"
15"
6"

1½" x 3½" NOTCH (3 TYPICAL)

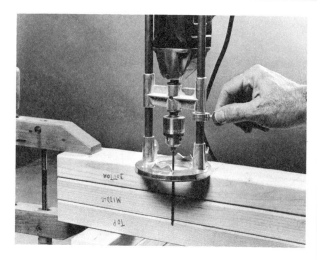

1. Clamp the three rail members together, then set the drill to penetrate only partly into the lowest rail in the stack (the top rail, bottom side up). Bore the pilot holes.

4. The radial-arm or table saw can be used to make the parallel cuts for the notches in the facing boards.

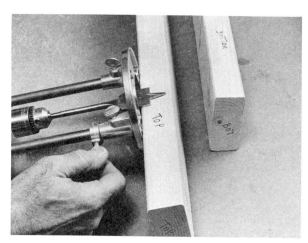

2. Adjust a flat bit to make a ½-inch-deep blind hole in the top and bottom rails.

3. Bore through holes in the middle rail only. The pre-drilled pilot holes ensure precise alignment.

5. Finish off the cuts with a sabre saw.

6. Use the post face board to hold the rails in position while inserting the pickets.

7. Remove, then reinsert the face board after the bottom rail has been added.

8. Check the assembly for squareness, then drive nails through the rails into each picket.

9. Ready for final assembly at the site. The 6 × 6 post would of course be pre-set in the ground. Toenail the rails into the post.

26. HOW TO BUILD A BETTER GATE

While sound construction is essential in a fence, a gate needs even greater structural strength because it naturally takes more abuse. Unless it is well built, constant swinging and slamming will soon cause it to sag or bind and eventually fall apart. Inadequate hinges or fasteners may also cause gate failure. The basic framing techniques detailed here will help you to build a sound, lasting gate.

The design of your gate will be dictated by the design of your fence. The gate can match the fence exactly, it can be designed to accentuate the entryway, or you may make design changes to reduce the gate's bulk and weight. Whatever your preference, one of the framing methods shown will very likely be adaptable.

One gate features a frame assembled with dowel-reinforced end rabbet joints and let-in plywood gussets. The other achieves strength and rigidity through the use of cross bracing joined to the rails with end and crossed half-lap joints. Waterproof

The arched pickets and ornamental hardware enhance the appearance of this gate. The crossed bracing adds strength.

This sturdy gate retains the lines of the fence but it is less bulky due to the use of lighter rails and elimination of the skirt.

This rear view shows how the let-in gussets add an interesting design element while strengthening the frame considerably. Extra-long strap hinges are less apt to sag.

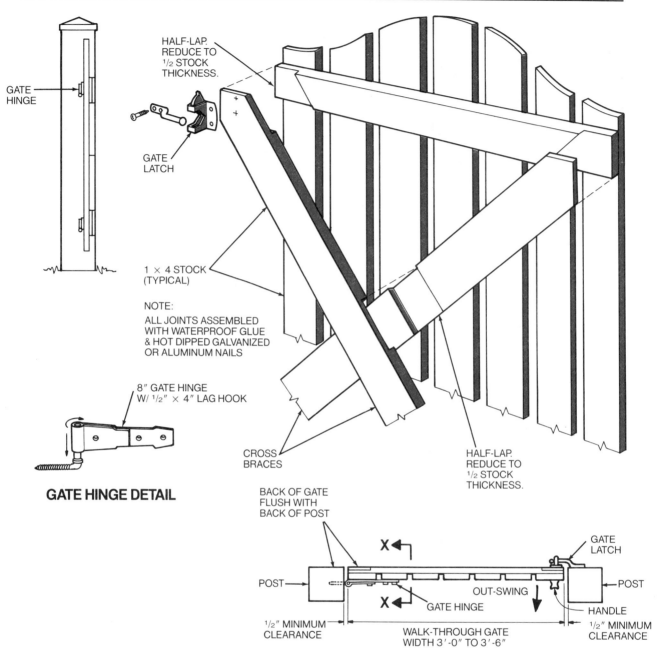

GATE HINGE

HALF-LAP. REDUCE TO 1/2 STOCK THICKNESS.

GATE LATCH

1 × 4 STOCK (TYPICAL)

NOTE:
ALL JOINTS ASSEMBLED WITH WATERPROOF GLUE & HOT DIPPED GALVANIZED OR ALUMINUM NAILS

8" GATE HINGE W/ 1/2" × 4" LAG HOOK

GATE HINGE DETAIL

CROSS BRACES

HALF-LAP. REDUCE TO 1/2 STOCK THICKNESS.

BACK OF GATE FLUSH WITH BACK OF POST

GATE LATCH

POST

OUT-SWING

POST

X

GATE HINGE

X

HANDLE

1/2" MINIMUM CLEARANCE

WALK-THROUGH GATE WIDTH 3'-0" TO 3'-6"

1/2" MINIMUM CLEARANCE

resorcinol resin glue is used in both framings. The combination of rabbeted corners and increased glue surface provided by these joints strengthens the frames.

The joints are not difficult to make. Use a table or radial-arm saw with a dado head to make the cuts. If you do not have one of these machines, you can do an excellent job with a portable circular saw and a simple jig, which can be made with two scraps of wood. The procedures involving both the stationary and portable saws are illustrated, so you can take your pick.

It obviously is necessary to design the gate before you begin to construct it, so there are some matters

to consider. If the fence has already been erected, the width of the gate will be determined by the space between the gateposts. If you are building a new fence, allow at least 3 to 3 1/2 feet for an average walk-through gate. If oversized garden equipment must be moved through the gate, adjust the width accordingly.

Assuming the fence is already erected, measure the space between the posts and size the gate so there will be at least 1/2 inch clearance on each side. If the cross section thickness of the proposed gate is relatively large, it is advisable to make a full-sized flat section mock-up of the posts and gate in order to determine the necessary swing clearance accu-

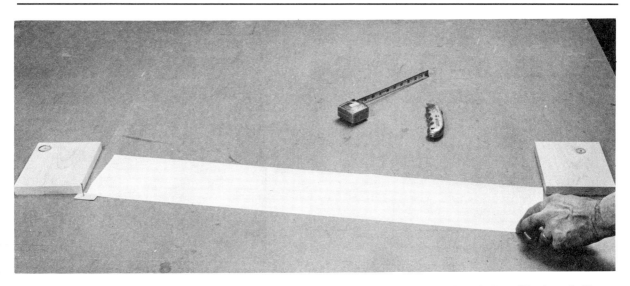

A full-sized mock-up of gate and posts helps to determine the proper swing clearance. Note how the "gate" is pivoted with a nail at the hinge-pin position.

You can give a gate with a thick cross section extra clearance simply by rounding the leading edge on the latch side.

rately. To do this, cut two pieces of wood or cardboard sized to represent the posts. Space them on the worktable accordingly. Cut a piece of cardboard to size to represent the top view section of the gate. Tape a projecting tab at the hinge location; then drive a nail through the tab at the estimated pivot position of the hinge. Swing the "gate" and observe the clearance. Alter the dimension if necessary and size the actual gate accordingly. Note that clearance can be gained by rounding the leading edge of the gate on the latch side. This can also be worked out on the mock-up. This simple trial pro-

cedure is also useful to determine how to space the gateposts if the fence is still unbuilt.

During the design stage, you will need to consider the desired swing of the gate; do you want it to swing inward or outward? Your decision will have a direct bearing on the design as well as the selection and positioning of compatible hardware. The hardware applications shown in these sample gates are typical. Note that the design of the heavy gate does not lend for proper out-swinging front-hinge application due to the lack of suitable flat mounting surfaces.

In order to improve the design of the gate for the heavier fence, we did the following: 2 × 3 framing was used rather than the 2 × 4 stock that was used for the fence; the bottom skirt was also eliminated. The result is a relatively light gate that is clearly distinguishable from the rest of the fence due to the absence of the skirt and the slight set-back afforded by the thinner cross section. A further advantage in eliminating the skirt, by the way, avoids the possibility that the gate will drag and snag on uneven ground.

The straight picket gate can be made more prominent simply by using longer arched pickets. When considering a design, it will prove helpful to make a scale drawing of a portion of the fence and the proposed gate. This will give you an opportunity to visualize the end result in advance. A scale drawing will also indicate the stock sizes required for the pickets and braces. Use 1/4-inch-grid graph paper to make the scale drawing, letting each square represent 1 inch, or use an architect's scale rule.

The step-by-step photos will show you how to frame and face a gate of your own design.

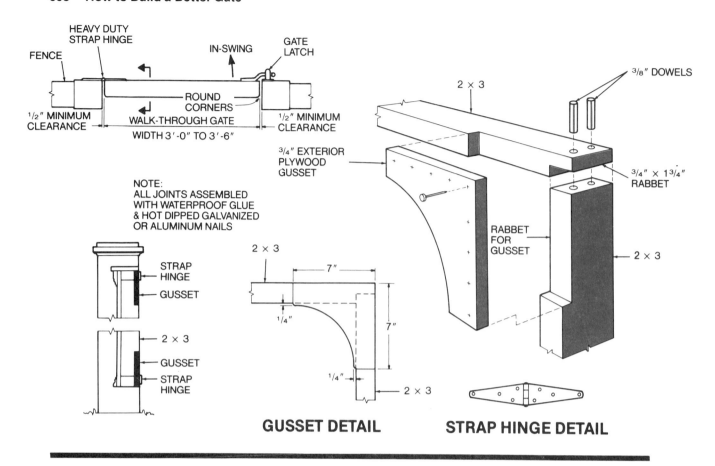

HEAVY DUTY
STRAP HINGE

IN-SWING

GATE
LATCH

FENCE

ROUND
CORNERS

1/2" MINIMUM
CLEARANCE

WALK-THROUGH GATE

1/2" MINIMUM
CLEARANCE

WIDTH 3'-0" TO 3'-6"

2 × 3

3/8" DOWELS

3/4" EXTERIOR
PLYWOOD
GUSSET

3/4" × 1 3/4"
RABBET

NOTE:
ALL JOINTS ASSEMBLED
WITH WATERPROOF GLUE
& HOT DIPPED GALVANIZED
OR ALUMINUM NAILS

RABBET
FOR
GUSSET

2 × 3

STRAP
HINGE

GUSSET

2 × 3

GUSSET

STRAP
HINGE

2 × 3

7"

1/4"

7"

1/4"

2 × 3

GUSSET DETAIL

STRAP HINGE DETAIL

Cutting Framed Gate Rabbets

1. Cut the framing members to length and assemble them in place to mark the cutting lines for the rabbets.

2. Cut the end rabbets and position the frame members. Then place the gussets over the corners to mark the border lines for the second rabbet cuts.

Making Gusset Corners

3. A ³/₄-inch-deep cut is made to form the recess for the gusset. The adjustable dado head is set for its widest cut.

4. Clean true cuts are essential for a strong glue joint. This corner treatment is suitable for many types of gates.

5. The frame is dry-assembled with bar clamps to facilitate drilling the through dowel holes. The tape marker on the bit serves as a depth gauge.

6. Use waterproof glue and spiral-grooved dowels to assemble the frame. Dowels are especially recommended here because nails do not hold well in the end grain of softwood.

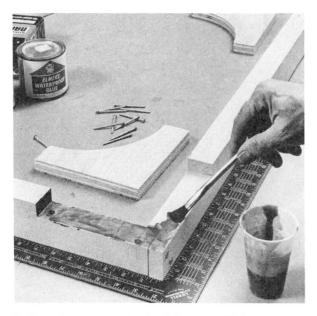 

7. Use a large square to check for a true 90-degree corner before attaching the gusset. This should be done before the glue in the doweled corners begins to set (so that minor adjustments can still be made).

8. Use only hot-dipped galvanized or aluminum nails for gate construction. The corner nail was positioned so that it missed the dowel.

9. The completed frame is trim but quite strong. It becomes even stronger when the pickets are added.

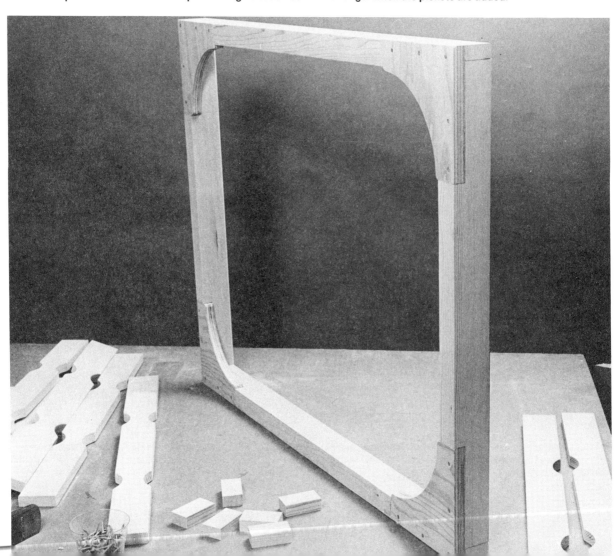

Jig for Scallop Cuts

10. Should your gate require numerous scallop cuts, this simple jig and a hole saw will save you much time and effort. Each cut produces a scallop in two pickets simultaneously. Oversized hole saws are readily available from Sears Roebuck.

Assembling the Gate

11. Filler blocks are used to seal the spaces between the pickets on the top rail. They can be left off the bottom rail to permit water runoff.

12. Set all nails and fill with putty before painting.

13. Ease all sharp outside corners of the gate to minimize risk of dangerous accidental body contact.

14. Fold the hinge back and butt it firmly against the side of the gate to obtain a true alignment. If the pins of the paired hinges are off axis, the gate will squeak and swing poorly.

15. A swivel caster provides good support for a heavy wide gate if the caster rides on level ground. Otherwise the caster stem must be spring-tensioned.

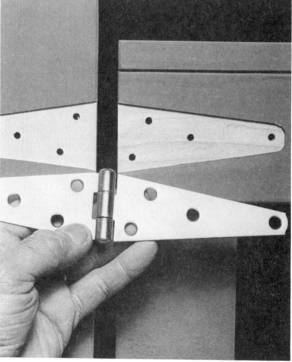

16. On very heavy gates, the hinges can be made sag-proof by setting them into mortises cut in the gate and post. Another method of sag-proofing is to use through bolts with nuts.

Laying out the Arched Picket Gate

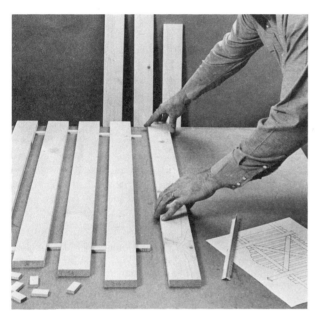

1. A scale drawing helps in sizing and laying out the components of the gate. Start by spacing the pickets with the aid of spacing blocks.

2. Use small nails to tack the upper and lower rails in place temporarily. Use a large square to true the assembly; then drive two nails into each corner to prevent shifting.

3. Lay in the first diagonal and mark the rails for the half-lap cuts. Also mark the projecting ends for cutoff. Before removing the brace piece, set the T bevel to the working angle.

4. Use the T bevel to carry the lap-joint lines to the bottom surface of the diagonal member.

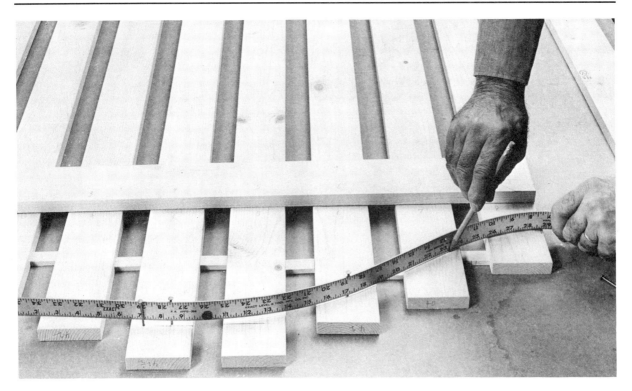

5. Three nails and a flexible rule are used to draw a graceful curve across the top of the pickets. Only half the pickets need to be marked in this way. The others can be marked by tracing from the cut pickets.

Cutting the Half-Lap Joints

6. Two crossed boards nailed together at the angle prescribed by the T bevel serve as the guide for cutting the half-lap joints with a portable saw.

7. Set the blade projection to penetrate halfway through the stock. The kerf cut in the jig is aligned with the marked joint line on the work. A temporary nail is driven into the workpiece to hold the jig in place only for the first cut.

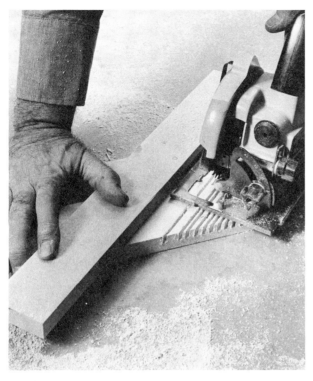

8. After the first cut has been made, the holding nail is removed and the jig is guided freehand for repeated kerf cuts.

9. A chisel is used to cut away the waste slivers flush to the surface of the cut.

10. When the end half-laps have been cut, the two brace pieces are marked in place for the cross half-laps. These angles will differ from the others, so a new jig must be made.

11. Check the pieces for fit after all the laps have been cut. Be sure to identify them for later matching.

Cutting the Arched Pickets

12. Use a sabre saw to cut the picket end curves.

Final Assembly

13. Apply waterproof glue and assemble the gate with non-rusting hot-dipped galvanized (or aluminum) nails.

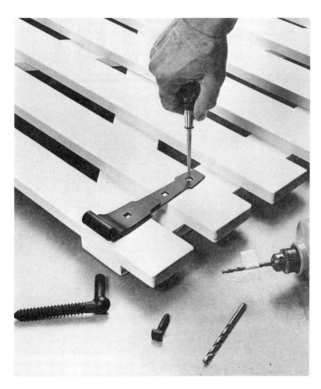

14. Attach the strap member of this hinge to the gate first so that you can measure the distance for spacing the lags on the post without difficulty.

15. The pilot hole for the hinge lag must be bored perfectly perpendicular if the gate is to hang true, so the use of a drill guide is altogether essential.

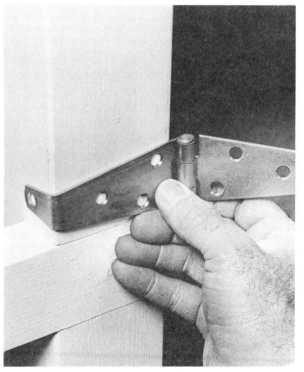

16. Lube the threads with wax or soap and turn the lag into place. A wrench will be necessary once the threads take hold.

Handy hint: when it is desirable to use an extra-long heavy-duty strap hinge on a 4 × 4 post, simply make a right-angle bend in the leaf so it will wrap around the corner of the post.

27. BARBECUE PANTRY

When you have a backyard barbecue, does it turn into a relay race to and from the kitchen? Build this handy pantry, store all your outdoor cooking equipment and condiments in it, and save the running for fun and games instead of step-and-go-fetch-it.

Simple lines make the unit a snap to build even if you've never constructed anything before. The dual-purpose panel covers and protects the counter top and stores utensils, and a lock can be added if desired. The counter top is flanked at both ends with built-in recesses. A deep one stores bottled condiments and a shallow one accommodates smaller seasoning containers. Toe space comes from slanting the front inward at the bottom, giving the piece attractive lines and eliminating a cutout at the bottom. Large casters can be added to allow mobility of the unit, but these are optional.

Weather-sealed tilt-out bins store supplies. The narrow one takes a 20-pound bag of charcoal and a can of starter fluid. The larger one stores cups, glasses, a coffee maker, paper goods, and sundries. These bins are easy on your back because you can reach their contents without stooping and bending your knees.

Build the cabinet of 1/2- and 3/4-inch exterior plywood. The design has been kept simple so you can build it with hand power tools exclusively, if necessary. The butt joints are assembled with finishing nails and waterproof glue. Be sure to use only hot-dipped galvanized nails, and set and fill the depressions.

Pressure-sensitive sponge-rubber weather-stripping tape applied to the inside edges of the bins and across the top of the back panel makes this pantry weathertight so it can be kept on the back porch or in some other convenient but not completely protected area.

When assembling the piece, the piano (strip) hinge pivot point must be offset 1/8 inch to favor the rubber strip. This allows 1/16 inch compression of the 3/16-inch-thick by 3/8-inch-wide weather stripping—enough to exclude water and dampness so the pantry can be used to store all the paper goods and gear needed for a quick and enjoyable cookout.

The two-tone final finish makes for an attractive appearance. A coat of wood sealer tames the fir grain before sanding and applying exterior enamel.

The watertight cabinet can be kept anyplace outdoors.

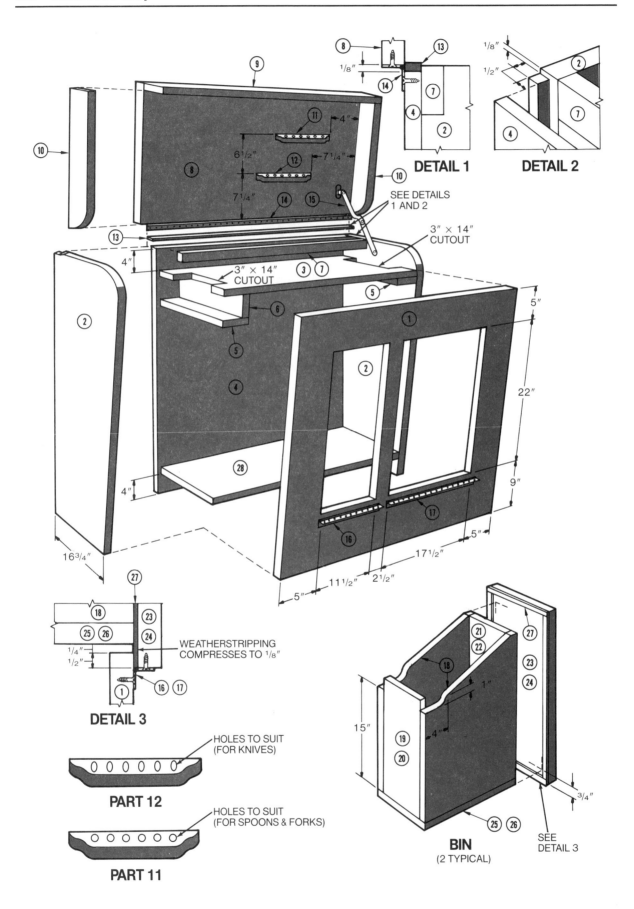

DETAIL 1

DETAIL 2

SEE DETAILS
1 AND 2

3″ × 14″
CUTOUT

3″ × 14″
CUTOUT

5″

22″

9″

5″

16³/₄″

17¹/₂″

5″

11¹/₂″

2¹/₂″

5″

WEATHERSTRIPPING
COMPRESSES TO ¹/₈″

¹/₄″
¹/₂″

DETAIL 3

HOLES TO SUIT
(FOR KNIVES)

PART 12

HOLES TO SUIT
(FOR SPOONS & FORKS)

PART 11

BIN
(2 TYPICAL)

15″

1″

4″

3/4″

SEE
DETAIL 3

Construction Tips

1. A guide strip tack nailed at an angle produces the slanted cut for both side panels at one time. Be sure to measure accurately so both pieces come out the same size.

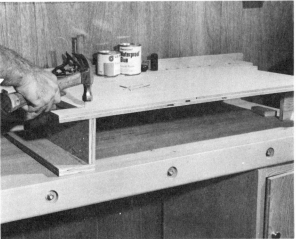

2. The recessed partition is built into the counter top before the cabinet is assembled.

BARBECUE PANTRY

MATERIALS LIST

ITEM	QTY	DESCRIPTION
1	1	3/4" × 36" × 41 1/2" plywood
2	2	3/4" × 20 1/4" × 40" plywood
3	1	3/4" × 19" × 40" plywood
4	1	1/2" × 40" × 40 1/2" plywood
5	2	3/4" × 4" × 19" plywood
6	1	3/4" × 4 1/2" × 19" plywood
7	1	3/4" × 1 1/2" × 40" plywood
8	1	3/4" × 21" × 43" plywood
9	1	3/4" × 4 1/2" × 43" plywood
10	2	3/4" × 4 1/2" × 20 1/4" plywood
11	1	3/4" × 1 1/2" × 11" plywood
12	1	3/4" × 1 1/2" × 11" plywood
13	1	3/16" × 1/2" × 41 1/2" weatherstripping
14	1	1 1/2" × 41 1/2" strip hinge
15	2	lid support
16	1	1 1/2" × 13" strip hinge
17	1	1 1/2" × 19" strip hinge
18	4	1/2" × 16" × 20" plywood
19	1	1/2" × 10" × 15" plywood
20	1	1/2" × 15" × 16" plywood
21	1	1/2" × 10" × 20" plywood
22	1	1/2" × 16" × 20" plywood
23	1	3/4" × 3 1/16" × 23" plywood
24	1	3/4" × 19" × 23" plywood
25	1	1/2" × 11" × 16" plywood
26	1	1/2" × 16" × 17" plywood
27		3/16" thick × 3/8" weatherstripping (13' required)
28	1	3/4" × 16 11/16" × 40" plywood

Note: all plywood to be exterior grade

3. Nailed butt joints and waterproof glue make for easy assembly. Note how the parallel clamp is used as a third hand.

4. The openings in the one-piece front panel are cut with a sabre saw. Drill corner holes first and use a clamped guide to make precise cuts.

5. A router with a rounding-over bit is used to shape the cabinet edges round. The bin front panel edges are only slightly rounded by sanding.

6. The hinge is attached to the bottom edge of the bin's front panel first, with an offset of ⅛ inch. The bin is then inserted in the cabinet to mark the hinge screw centers in the front panel. This is done before the weather stripping is attached.

7. Self-adhering 3/16 by 3/8 inch weather stripping is easy to apply. Just peel off the protective strip and roll the band into place. The areas to be treated with weather stripping should be pre-painted.

28. SWINGING STORAGE CABINET

This unique stack of swing-out trays offers handy storage for numerous items in the home, shop, or office. A flick of the finger is all it takes to reach the contents, and the unit rolls about on casters for added convenience.

The trays are made of ¼-inch fir plywood and assembled with glue and brads. A solid lumber block installed in the corner of each tray mounts over a length of thin-wall electrical conduit that serves as a through pivot.

Use a smooth cutting blade, preferably hollow ground, to make the parts for the trays. This will save much time and effort in sanding the edges satin smooth. Before assembly sand all the inside surfaces, but be careful to avoid a "cushion" effect near the ends. Since the relatively thin stock provides little surface glue contact, it is important to have well-fitted butt joints. To make these, it is advisable to sand by hand, with a wood block back-up. You should not use a finishing sander because it will cause some slight rounding.

This project will give you plenty of experience in driving nails into the edges of narrow stock. Use ¾-inch 20-gauge brads and work with a light 7-ounce hammer. Start two brads into a joint to obtain registration holes before you apply a bead of glue. When you have assembled the trays, you can then sand the outsides. A stationary belt sander is OK for the broad surfaces, but use a finishing sander to smooth the top edges and to ease all the sharp corners.

Use a stable wood of medium hardness, such as poplar, for the tray corner blocks. Dress a length of ⁸/₄ square stock; then cut a bevel on one corner and sand the bevel and the two adjoining flats. Do not sand the surfaces that will be glued. Next, use a block plane to shave a slight bevel of about ⅛ inch on the back corner. This will assure a good fit of the

blocks into the tray corners. Now make cross cuts to slice off nine blocks 2¾ inches long. Sand the top ends of each block very smooth so that they will not cause friction problems in the finished unit. Glue the blocks into the corners of the trays.

The holes in the blocks must be bored perfectly perpendicular and in exact alignment if they are to line up neatly and swing out accurately. You can do

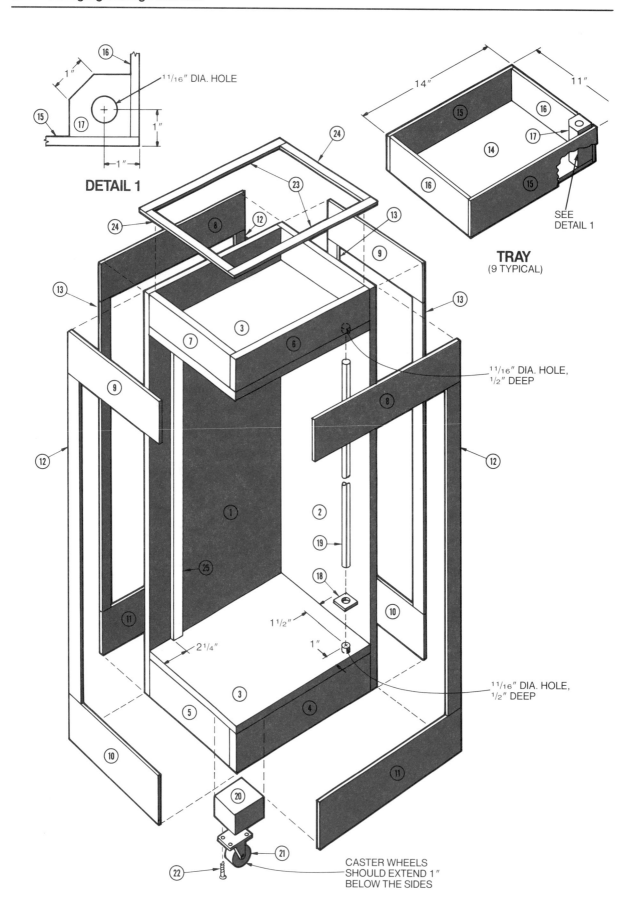

DETAIL 1

$^{11}/_{16}''$ DIA. HOLE

1″

1″

1″

TRAY
(9 TYPICAL)

SEE
DETAIL 1

14″

11″

$^{11}/_{16}''$ DIA. HOLE,
$^{1}/_{2}''$ DEEP

$^{11}/_{16}''$ DIA. HOLE,
$^{1}/_{2}''$ DEEP

$1^{1}/_{2}''$

$2^{1}/_{4}''$

1″

CASTER WHEELS
SHOULD EXTEND 1″
BELOW THE SIDES

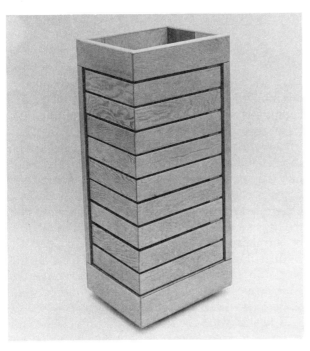

The type of wood you use is optional. Fir plywood is the best choice if cost is a major factor. If a finer look is desired, any fancy hardwood veneer plywood can be used.

The nested trays will respond nicely to fingertip pressure if the pivot blocks are waxed and well fitted.

SWINGING STORAGE CABINET

MATERIALS LIST

ITEM	QTY	DESCRIPTION
1	1	$3/4'' \times 14^1/2'' \times 34^9/16''$ plywood
2	1	$3/4'' \times 12^3/16'' \times 34^9/16''$ plywood
3	2	$3/4'' \times 11^7/16'' \times 14^1/2''$ plywood
4	1	$3/4'' \times 3^3/16'' \times 14^1/2''$ plywood
5	1	$3/4'' \times 3^3/16'' \times 10^{11}/16''$ plywood
6	1	$3/4'' \times 2^3/8'' \times 14^1/2''$ plywood
7	1	$3/4'' \times 2^3/8'' \times 10^{11}/16''$ plywood
8	2	$1/4'' \times 3^3/8'' \times 15^3/4''$ plywood
9	2	$1/4'' \times 3^3/8'' \times 12^3/16''$ plywood
10	2	$1/4'' \times 3^{15}/16'' \times 12^3/16''$ plywood
11	2	$1/4'' \times 3^{15}/16'' \times 15^3/4''$ plywood
12	3	$1/4'' \times 1^1/4'' \times 27^1/4''$ plywood
13	3	$1/4'' \times 1'' \times 27^1/4''$ plywood
14	9	$1/4'' \times 10^1/2'' \times 13^1/2''$ plywood
15	18	$1/4'' \times 2^3/4'' \times 14''$ plywood
16	18	$1/4'' \times 2^3/4'' \times 10^1/2''$ plywood
17	9	$1^3/4'' \times 1^3/4'' \times 2^3/4''$ poplar
18	1	$1/4'' \times 1^3/4'' \times 1^3/4''$ plywood
19	1	$1/2''$ thin wall conduit, $28^1/4''$ long
20	4	$2^3/4'' \times 2^3/4'' \times 2^1/4''$ high block
21	4	swivel plate caster, $1^5/8''$ wheel
22	16	$1^1/4''$ # 10 rh wood screw
23	2	$1/4'' \times 1'' \times 15^3/4''$ solid pine
24	2	$1/4'' \times 1'' \times 10^{11}/16''$ solid pine
25	1	$1/2'' \times 1'' \times 27^1/4''$ pine

this easily if you use a simple line-up guide. Place a scrap board, a bit larger than the tray, on the drill-press table. Align the center marked block under the drill bit; then clamp two stop blocks at the corner of the tray. Hold each tray firmly against the stops while boring the holes.

The size of the conduit used for the pivot is $1/2$ inch, with an outside diameter of $^{11}/_{16}$ inch. Use an $^{11}/_{16}$-inch flat bit to make the holes. Check the fit of the conduit in a hole. If it is too snug, you will need to do some sanding to obtain a good slip fit. Sanding the inside of the hole in each block will be somewhat time consuming and difficult, so it is preferable to sand the conduit. Start with 80-grit aluminum oxide paper backed with a padded block. When the size is right, finish off first with 100-grit, then 120-grit paper.

When you have bored all the holes, apply a coat of sanding sealer to all the surfaces of the trays, including the bottoms and particularly inside the holes. Using 220-grit paper, lightly sand the walls of the holes and all the surfaces of the trays. Wipe off the sanding dust; then rub a wax candle over the walls of the holes and to the top of the pivot blocks for lubrication. Slide the trays over the conduit.

Make the cabinet of $3/4$-inch fir plywood. Drill a blind hole $1/2$ inch deep into the top of the base panel and into the bottom of the upper panel. Glue and nail the two cabinet side panels together. Pre-

Making Drawers

assemble the upper and lower sections, which are comprised of three pieces each. Glue and nail the lower section into place.

Carry the conduit with the trays attached to the cabinet and insert the protruding end of the conduit into the hole in the base. Place the still-unattached upper section over the other end of the conduit. Carefully keep everything intact, and glue and nail the upper section into place. The conduit will now be permanently installed.

Turn the unit bottom side up and install the blocks in each corner of the base to receive the plate-swivel casters. These blocks should be sized so that the wheels project about one inch.

A facing of 1/4-inch plywood is attached to all exposed surfaces except for the top edges of the upper compartment, where a facing of 1/4-by-1-inch solid pine is used. Ease all corners with a block plane and sand. A 1/2-by-1-inch strip of pine is attached near the front edge of the inside wall to serve as a stop for the trays. You will find that the wild grain pattern of fir plywood can add an interesting effect when finished natural. Two coats of Constantine's Satin Finish Wood-Glo were used on the prototype. An alternative finish could be a two-tone enamel.

1. Glue and 3/4-inch 20-gauge brads are used to assemble the trays. Flat, unwarped panels are essential.

2. Ease all sharp corners with a block plane; then sand. Finish-sand the inside surfaces before assembly.

3. The glued corner blocks are easily held in place with a single C clamp. A V-notched block cut from scrap bridges the outside corner.

Aligning Holes

4. Precise alignment of all the corner post holes is important. Use corner stops like this one for consistency.

5. The insides of the holes are given a coat of sealer; then they are sanded lightly. A wad of cotton on a stick makes application of the sealer easy.

6. The trays are stacked over a length of thin-wall conduit. A coating of wax inside the holes and on top of the blocks will ensure free movement.

Cabinet Construction

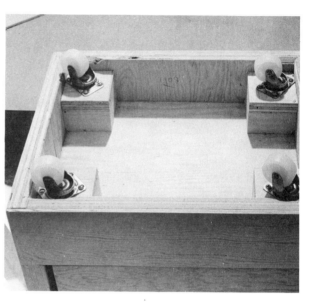

7. The upper and lower sections have blind holes to receive the conduit. One section *only*, either one, is attached to the cabinet sides in advance.

9. Swivel-plate casters are attached to blocks which are glued into each corner of the base.

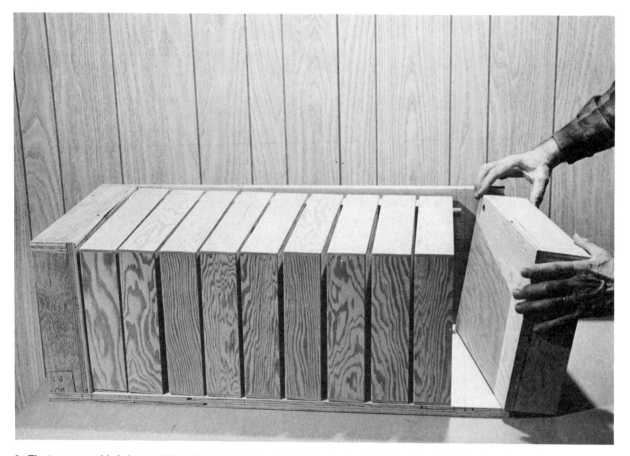

8. The tray assembly is inserted into the cabinet as a unit and then locked into place by attaching the other end section.

29. WALL WINE RACK

This charming wine rack cradles five bottles in the traditional prone position. The bottles are easily accessible, and their labels are clearly displayed for quick identification. You can easily alter the length of the rack so that it can hold up to ten or even a dozen bottles. You need only to repeat the pattern of the side panels to stretch them out as desired.

Use any species of kiln-dried hardwood. The one shown here is made of cherry—an ideal choice for this project because it tools and sands exceptionally well. This is an important factor due to the abundance of exposed end grain, which is normally somewhat more difficult to cut and smooth properly than is face grain. The four parts required can be cut out of a 1-by-6-inch-by-6-foot board.

The curvy configurations on the side pieces can be cut with ease and accuracy if you first bore a series of holes as indicated in the drawing. Use a 1½-inch flat bit and a fly cutter adjusted to a 3⅜-inch diameter. This is a job for the drill press. *Never* attempt to use a fly cutter in a portable drill because they're not made for each other. To do so would surely result in serious bodily injury.

After the holes are made, it is a simple matter to cut out the waste by sawing on the lines tangent to the circles. Smooth out the saw ripples by sanding with a drum sander chucked in the drill press.

Use a router with a sharp ³⁄₈-inch corner-rounding bit to shape the edges. A dull bit may tend to chip out the wood at the narrow ends of the "fingers."

A pair of ³⁄₈-inch dowels are used to join each side to the top and bottom back pieces. Drill the holes in the sides first; then insert a pair of dowel center pins to transfer and mark the holes accurately in the edges of the back pieces. Glue and clamp the sides to the backs.

The best and safest way to hang the rack is with a screw driven directly into a wall stud. If the rack

329

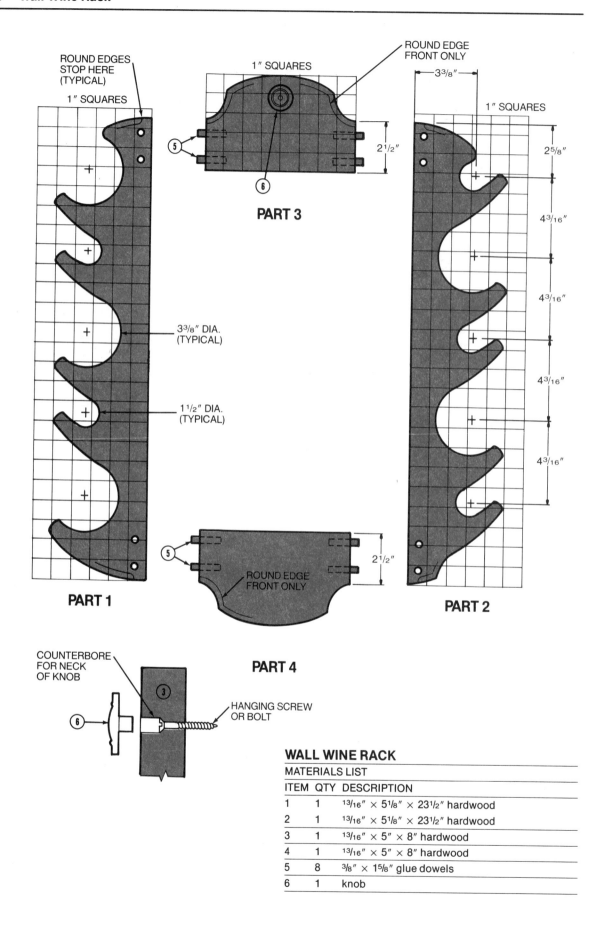

ROUND EDGES
STOP HERE
(TYPICAL)

1″ SQUARES

ROUND EDGE
FRONT ONLY

1″ SQUARES

3³/₈″

1″ SQUARES

2¹/₂″

PART 3

2⁵/₈″

4³/₁₆″

4³/₁₆″

3³/₈″ DIA.
(TYPICAL)

4³/₁₆″

1¹/₂″ DIA.
(TYPICAL)

4³/₁₆″

4³/₁₆″

PART 1

2¹/₂″

ROUND EDGE
FRONT ONLY

PART 2

COUNTERBORE
FOR NECK
OF KNOB

PART 4

HANGING SCREW
OR BOLT

WALL WINE RACK

MATERIALS LIST

ITEM	QTY	DESCRIPTION
1	1	¹³/₁₆″ × 5¹/₈″ × 23¹/₂″ hardwood
2	1	¹³/₁₆″ × 5¹/₈″ × 23¹/₂″ hardwood
3	1	¹³/₁₆″ × 5″ × 8″ hardwood
4	1	¹³/₁₆″ × 5″ × 8″ hardwood
5	8	³/₈″ × 1⁵/₈″ glue dowels
6	1	knob

must be hung on a hollow wall, use a toggle bolt or an expansion anchor. A decorative cabinet knob is used to conceal the securing screw or bolt. Measure the diameter of the neck of the knob and counterbore a hole of this size about a third of the way into the back. Cut off part of the neck of the knob, if necessary, so that it will fit flush. Apply a coat of sanding sealer to the rack, then two coats of self-rubbing clear finish. Sand with #600 paper between coats, but leave the final coat unsanded.

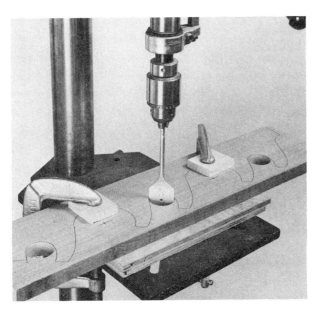

1. A flat bit of this size should be driven at medium speed for hardwood. Too high a speed will overheat the steel, draw the temper, and make it dull. Too low a speed will result in a rough cut. Be sure to use a back-up board to prevent splintering on the exit side (and also to prevent the bit from touching the table).

2. Use a fly cutter to make the large holes. Bore halfway through; then turn the piece over and complete the cut from the other side. The pilot-bit hole will ensure alignment. This cutter has great thrust, so the work must be securely clamped.

3. The band saw is used to drop out the waste. Follow the lines carefully so the cuts are tangent to the circles. This method is better than sawing the entire outline.

4. A 3/8-inch corner-rounding bit is used to shape the edges. All saw ripples must be sanded out before this step. If the bit is sharp, only minimal additional sanding will be required.

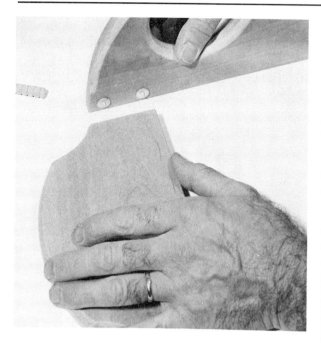

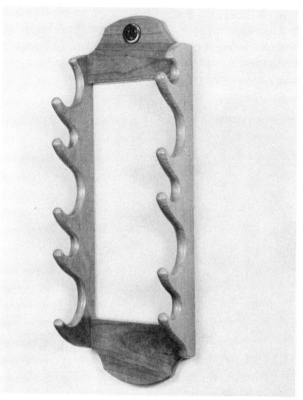

5. Dowel centers are used to obtain accurate drill centering marks. Insert the pins in the holes drilled in the side member, line up the parts, and press together. The center points will transfer to the second piece.

Detail view of the completed rack.

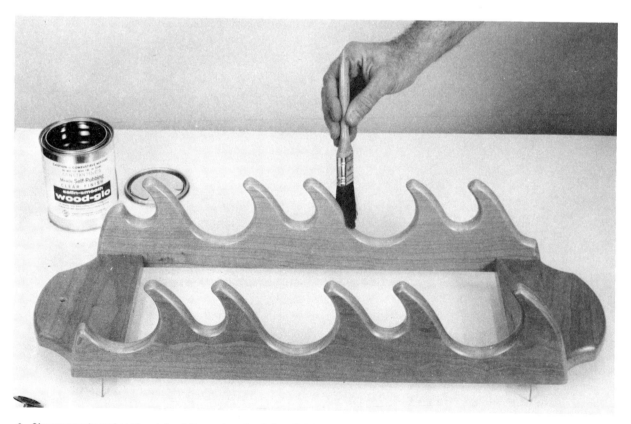

6. Cherry wood need not be stained. Several coats of clear finish will bring out its rich coloring and figure.

Appendix

A first-aid kit is a vital item in any shop. Keep one within easy reach and be sure to attend to any injury quickly, regardless of how minor it may seem.

SHOP SAFETY

In a sense, sharp-edged hand tools and fast-moving power machines are dumb: they don't know the difference between wood and flesh! Put a finger or two where only the wood should be, and you'll soon find out.

Many activities in the workshop are potentially dangerous, but you can avoid accidents and injuries if you develop safe and sound work habits and never let up your guard. Think and practice safety to the point that it becomes an unconscious, unbreakable habit. Don't be afraid of working with tools and machines, but do respect them to the fullest. Remember—they can't think, but you can.

Some basic "do" and "don't" rules follow. You'll note that they are merely matters of plain common sense, which is the key to safety.

Safety Rules

1. Know your power tool. Read the instruction manual carefully. Become thoroughly familiar with the tool's applications, limitations, and potential hazards.

2. Ground all tools. Make sure you follow all wiring codes and recommended electrical connection instructions.

3. Make all adjustments and cutter changeovers with the power off. Be sure to remove adjusting keys and wrenches before you turn on the power.

4. Make the shop child-proof with locks or master switches or by removing starter keys.

5. Wear proper apparel. Remove your tie, watch, and any jewelry. Tie back long hair. Use safety goggles or a face mask for operations that produce flying chips, dust, or sparks.

6. Avoid forcing a tool. It will perform better and more safely when it is used at the rate for which it was designed.

7. Select the correct tool for the job and use it in the manner for which it was designed.

8. Avoid drawing any sharp-edged tool towards your body.

9. Secure the work in a vise or with clamps or special holders whenever feasible.

10. Maintain tools carefully. Clean and lubricate them on a regular basis. Continually check for any faults that may develop. Always keep sharp-edged tools sharp; dull tools can cause accidents.

11. Use blade guards whenever feasible and use push sticks and hold downs to avoid getting the fingers too close to the blade.

12. Never guide an uneven edge against the rip fence of a table saw. Never allow the free end of a piece of cut stock between the fence and blade.

13. Wait until a blade comes to a full stop before reaching to remove scrap pieces from the table.

14. Keep the work area well lit and clean. Avoid having wood scraps and extension cables underfoot.

15. Use only recommended accessories and follow their accompanying instructions.

16. If a circuit breaker opens or a power failure occurs when you are operating a power tool, be sure to switch it off immediately.

17. Use flammable materials only in well-ventilated areas.

18. Never work with tools when you are over-tired or under the influence of alcohol.

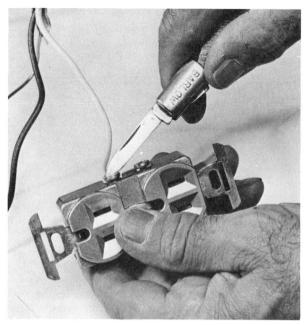

Use the right tool for the job. This is a good example of a bad practice. The tool is certain to be damaged, and if it slips the user's hand may be damaged as well.

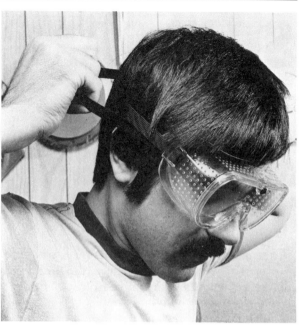

Use goggles for any operation that produces flying chips or grit. If you normally wear eyeglasses, make sure the lenses are shatterproof.

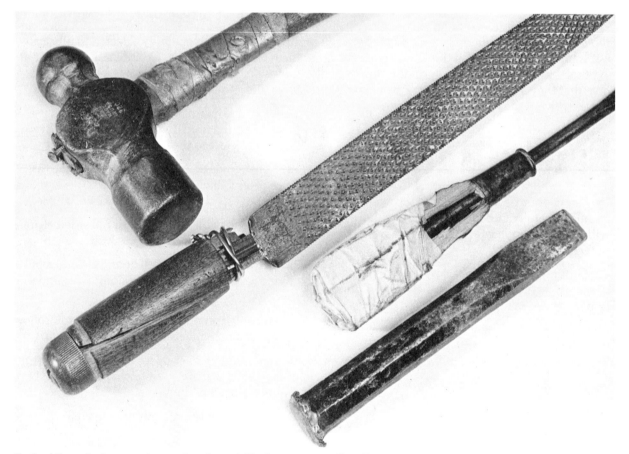

Each of these tools presents a serious hazard. The hammer and rasp handles should be replaced. The screwdriver should simply be discarded. A cold chisel head in this condition can easily splinter and possibly damage an eye in the event of a glancing blow. When a chisel head begins to fray, the edge should be ground down.

Don't hold small parts with your bare hands when using the grinding wheel. When this piece makes contact with the wheel it surely will go flying.

This is a better approach. Note that the see-through safety shield has been removed from the grinder in these situations for photo clarity.

Never use the band saw with the blade exposed above the work. A minor distraction here could result in quite a mess.

This is the only way to use the band saw—with the blade guard properly positioned.

This is a bad scene. The blade is too high and the fingers are too close for comfort. Also, the work should not be pushed on two sides of the blade.

This is better but not perfect—anti-kickback fingers should be used. A low blade and a push stick are correct however.

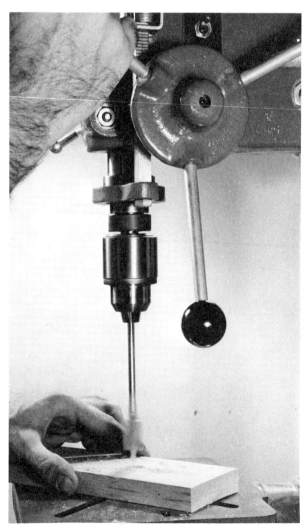

Wrong. Don't hold the work close to a drill bit.

Right. Clamp the work instead. Besides being safe, this will result in a cleaner cut.

Wrong. Never draw a sharp-edged tool toward the body.

Right. The cutting motion should be away from the body.

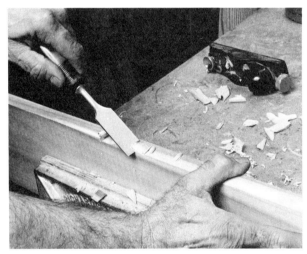

The thumb in the foreground is serving no purpose except to invite trouble.

This is the correct way to use a chisel. It is a tool that requires both hands.

SPECIALTY HARDWARE

A wide variety of hardware items are used in woodworking projects. Some are common and are readily available at local sources including hardware stores, lumberyards, and home centers. Others of a more specialized nature are generally not available from these sources, but they can be obtained from woodworkers' mail-order supply outlets. It pays to have their catalogs on file. Their selections will include special items that are required for many projects. And equally important, some pieces of special hardware may very well form the basis for project designs of your own. A sampling follows.

Cabinet Stays

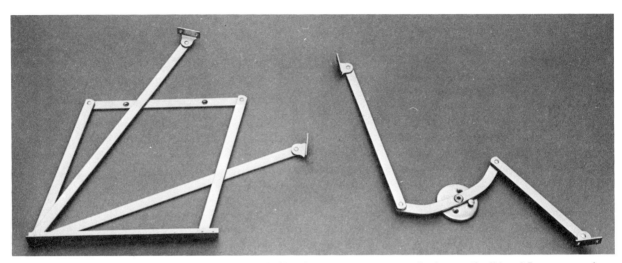

Unique cabinet fittings include triple-action bar stay (*left*) and double-action cocktail-cabinet stay. Both automatically raise the top lid when the front flap is lowered. The triple-action fix-ture goes one step farther: as the lid and flap open, a glass-ware tray is simultaneously lifted forward and upward toward the user. The mock-up illustrates these actions.

Triple-action stay, when closed glass tray is fully retracted.

As the front flap is lowered, the top lid pushes upward and the tray moves forward. When fully opened, the tray is positioned up front for easy access.

Cocktail-cabinet fitting in closed position . . .   . . . and fully open.

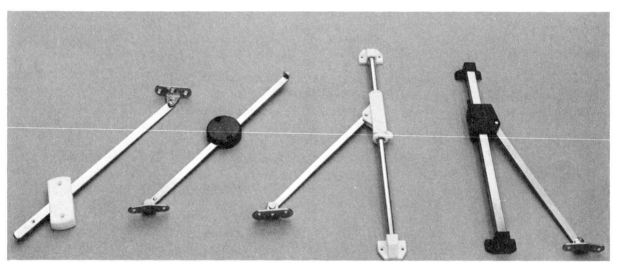

New and better drop-lid and lift-up flap supports use metal and plastic in a sliding action.

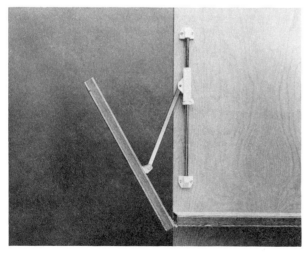

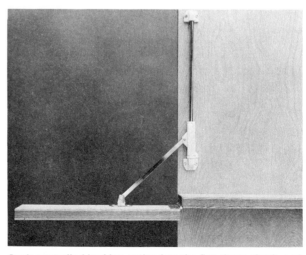

The friction ring on this unit can be adjusted to allow the flap to drop at a controlled rate.

Such controlled braking action lets the flap down slowly and silently; this action also prevents slamming and the tearing out of hinge screws.

Sliding Fixtures

TV extension fixtures solve the problem of obscured sight lines. Mounted on an extension chassis equipped with a swivel, the set can be moved out from its ordinarily fixed position, thus permitting viewing from just about anywhere in a 180-degree arc.

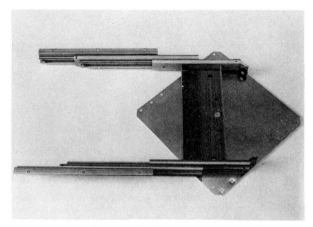

Light-duty swivels are available in sizes from 3 to 12 inches. They can be used for rotating servers, bookracks, taperacks, and the like.

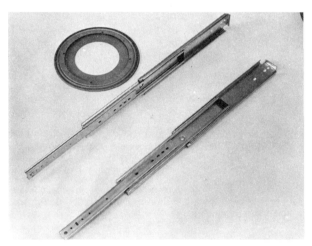

This unit features an integral swivel base.

Heavy-duty swivels are available for chairs, stools, and tables. This one has a memory: it turns a full 110 degrees to either side and returns automatically to the center.

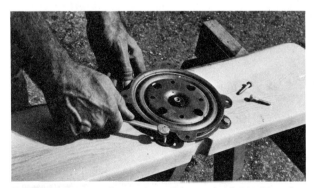

Full-extension extra-heavy-duty slides and a flat ball-bearing swivel can be combined for a custom-made chassis.

The Loc Trak system uses a plastic track that is inserted into a 1/8- by 3/8-inch deep saw kerf. A pair of nylon slides recessed into the door bottom ride smoothly over the track.

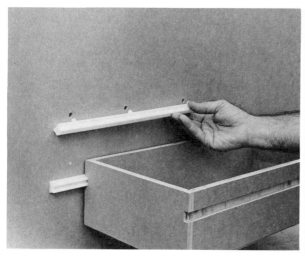

Drive-in plastic drawer runners are installed without screws. Installation simply requires drilling three holes. Grooves that have been cut into the sides of the drawer ride smoothly over the slippery runners.

A slide-out clothes hanger attaches to the bottom of a closet shelf and extends for easy access.

Concealed casters, practically invisible, mount on the inside walls of furniture or cabinets. They are available in both fixed and swivel types.

Hinges

Non-mortise butt hinges simplify the hanging of cabinet doors. The relatively thin leaves nest with each other, thus leaving a normal narrow gap between the door and frame.

Detail view of the novel hinge that requires no mortising.

The rising butt is an ingenious hinge that automatically lifts a door over thick carpeting or uneven flooring, eliminating jamming or scraping.

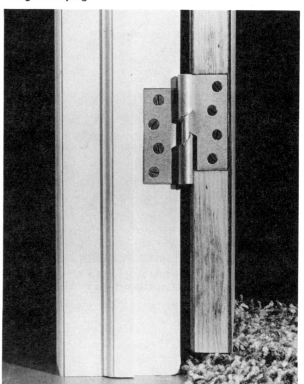

270° Hinge

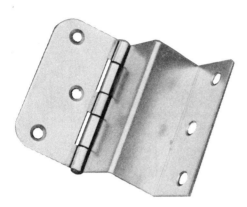

A swing-leg hinge rotates 270 degrees, providing a cabinet with a fold-out table that contains concealed legs.

This mock-up of a cupboard with fold-up table shows how the swing-leg hinge functions. When the table is folded up, the legs are inside and out of view.

The legs are swung in a 270-degree arc as the table is brought down from the cabinet.

The top edge of the legs tucks under the table to support it when horizontal.

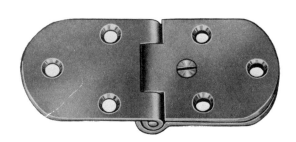

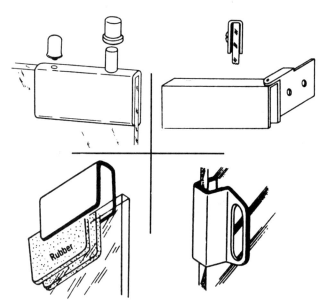

Butler-tray hinges are essential for butler-tray tables. They are spring-tensioned and lock precisely at 90 degrees.

Solid brass icebox hardware is available so that you can reproduce the old-fashioned oak icebox.

Glass-door hardware includes (*clockwise*) a self-locking hinge for recessed doors, a hinge for side wall mounting, a push-on strike plate for use with magnetic catch, and a slip-on door handle. None requires drilling of the glass.

Table Fixtures

Extension table slides feature rack-and-pinion glides. They are used for pull-apart and drop-leaf tables.

Table-top eveners are used to level divided tables.

An extension table lock keeps the sections from accidentally pulling apart.

Corner braces are used to reinforce tables and chairs. The angled ends are inserted into kerf cuts made in the apron. A hanger bolt with nut is then used to tighten the assembly.

Fasteners

Heavy-gauge steel bed fasteners are handy components for homemade beds. (*Above*) Wedge-action corner connectors permit instant assembly and knock-down of large furniture constructions. (*Below*)

Threaded inserts provide ¼-20 machine threads in wood. One end is slotted to take a screwdriver tip. The sharp outer threads easily screw into softwood, but threads should be tapped into hardwood.

Dot fasteners are an effective way to secure canvas to wood.

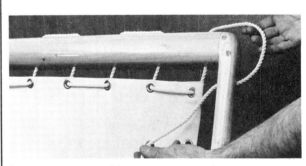

Brass grommets provide neat, strong eyes in canvas.

The pronged section is driven through the canvas against a softwood block after a center hole has been punched in the canvas. Prongs are then bent over a mating cap. Screw studs go into the wood.

A grommet setting die is relatively inexpensive.

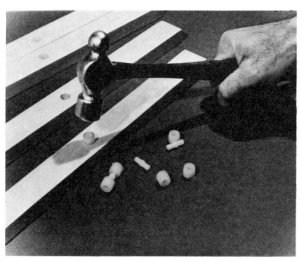

Nylon knock-down fittings are used for easy joining and detaching of smaller constructions. Bushings are press-fitted into ½-inch holes.

A double-sided dowel connects both members permanently or temporarily.

The fittings are used effectively in this adjustable screen project.

Sundries

This functional brass porthole is a novel accessory for the woodworker who is interested in constructing projects with a nautical motif.

An ornamental brass anchor is another item that may be of interest to the nautically inclined.

A brass spire finial will add a touch of elegance to any clock case or pediment.

A horsehead hook adds an interesting touch to a coat rack.

Turn-of-the-century-style barbershop coat hooks are becoming increasingly popular. This is the one shown in the Hall Butler project.

Finely detailed decorative overlay parts are available in both wood composition materials and solid brass.

Clock parts, including a wide variety of dials and movements, are available for the do-it-yourselfer interested in clock construction.

Weather-station instruments and ornamental figures make for interesting projects.

INDEX

INDEX